SpringerBriefs in Cancer Research

More information about this series at http://www.springer.com/series/10786

SpringerBriefs in Cancer Research

David H. Nguyen

Systems Biology of Tumor Physiology

Rethinking the Past, Defining the Future

 Springer

David H. Nguyen
Biological Systems & Engineering Division
Lawrence Berkeley National Lab
Berkeley, CA, USA

SpringerBriefs in Cancer Research
ISBN 978-3-319-25599-6   ISBN 978-3-319-25601-6   (eBook)
DOI 10.1007/978-3-319-25601-6

Library of Congress Control Number: 2015956773

Springer Cham Heidelberg New York Dordrecht London

Printed on acid-free paper

Springer International Publishing AG Switzerland is part of Springer Science+Business Media (www.springer.com)

*This book is dedicated to my friend
Caryn Mah, without whose encouragement
and support this book would have taken
much longer to complete. Thank you for
all the stimulating conversations about
radiologic imaging and for teaching
me that daffodils were not meant
to be studied under microscopes,
but that they were meant to be planted,
tended to, and delighted over.*

    *This book is also dedicated to all the
research technicians, research associates,
graduate students, postdoctoral fellows,
scientists, and principal investigators
who have contributed to my understanding
of mouse models in cancer research.*

Abstract

Systems biology can be defined as the study of how multiple interacting parts of a living system affect each other. A tumor is a system of cells that is imbedded in a larger system of organs, otherwise known as the human body. This book attempts three things from the systems biology perspective. The first chapter lays out testable reasons why estrogens and macrophages may be mechanisms behind why a lifelong exposure to estrogen increases breast cancer risk. The second chapter clarifies confusion between the cellular plasticity and cell-of-origin hypothesis in describing how tumors arise and how tumors sustain themselves. The third chapter describes common misuses of mouse models of cancer and how a systems biology perspective can extract so much more useful information from mouse models of cancer.

Preface

Systems biology can be defined as the study of how multiple interacting parts of a living system affect each other. A tumor is a system of cells that is imbedded in a larger system of organs, otherwise known as the human body. This book attempts three things from the systems biology perspective. The first chapter lays out testable reasons why estrogens and macrophages may be mechanisms behind why a lifelong exposure to estrogen increases breast cancer risk. The second chapter clarifies confusion between the cellular plasticity and cell-of-origin hypothesis in describing how tumors arise and how tumors sustain themselves. The third chapter describes common misuses of mouse models of cancer and how a systems biology perspective can extract so much more useful information from mouse models of cancer.

The primary purpose of this book is to provide insightful knowledge and testable predictions for researchers and students within cancer biology. The secondary goal is to provide training material for those who are new to the field of cancer biology. The first chapter will be very informative for researchers who study breast cancer risk and biological mechanisms of that risk. It will also be very useful to researchers who study cellular or molecular aspects of breast cancer and who want to better understand how global physiological mechanisms of normal mammary development and function affect their question of interest. The second chapter provides an example of how to assess two competing theories of tumor development. The third chapter provides concise, insightful guidelines that will bring newcomers to the topic of mouse models of cancer up to speed regarding common pitfalls and ways around them. It will also help veterans of mouse models of cancer to better make sense of their data: past, present, and future.

Berkeley, CA, USA   David H. Nguyen

Contents

**1 Macrophages, Extracellular Matrix, and Estrogens
in Breast Cancer Risk** 1

Introduction 1

Section 1: Estrus Cycle Regulates Macrophages
in the Mammary Gland 2

Section 2: The Estrus Cycle Regulates Macrophages in the Ovaries 3

Section 3: The Effects of Estrogens on Macrophages 4

Section 4: Macrophages as Targets of Endocrine Disrupting Chemicals 5

Section 5: Estrogen Effects on Macrophage ROS Production 6

Section 6: Complex Interplay Between Macrophages
and the Extracellular Matrix 7

    Composition of the Extracellular Matrix 7

    Macrophages Modify the Mammary Extracellular Matrix 7

    Effect of Increased Matrix Stiffness on Breast Cancer Progression 9

    Effect of Increased Matrix Stiffness on Macrophage Phenotypes 9

Section 7: Testing the Estrus-Regulated Macrophage-Dependent
Hypothesis of Breast Cancer Risk 10

    Directly and Indirectly Testing the Hypothesis 10

    Experimental Approaches to Directly Testing the Hypothesis 11

    Indirectly Testing the Hypothesis by Making Verifiable Predictions 12

    Closing Thoughts 15

References 16

2 Cellular Plasticity, Cancer Stem Cells, and Cells-of-Origin 21

Introduction 21

Section 1: Defining the Terminology Necessary
for Understanding This Field 22

    Stem Cell 22

    Progenitor Cell 22

    Differentiation 23

    Potency 23

Lineage Commitment.. 23
Transdifferentiation... 24
Dedifferentiation 24
Plasticity... 24
Epithelial-to-Mesenchymal Transition (EMT) 24
Niche 25
Microenvironment.. 25
Clonality... 26
Cancer Stem Cell 26
Cell-of-Origin 26
Section 2: Confusion Among Stem Cell, Cancer Stem Cell,
and Cell-of-Origin Concepts.. 26
Section 3: Evidence for the Theory of Cell-of-Origin
as the Root of Tumors 27
Section 4: Evidence for the Theory of Cellular Plasticity
in Tumor Development 28
Section 5: Confounding Factors in the Plasticity Versus
Cell-of-Origin Debate 28
Cell Fusion 29
Induced Mesenchymal States.. 29
References.. 30

3   Using Mouse Models and Making Sense of Them 33
Introduction.. 33
Section 1: Selecting the Right Model 35
The Age of the Mice 35
Appreciate the Estrus Cycle to Reduce Variation in Data 35
Background Genetics 37
Developmental Deformities Give Insight into Future Phenotypes 39
The Composition of Mouse Chow 40
Section 2: Correctly Using the Right Model.. 40
Mock Procedures for Control Groups Yield Cleaner
and More Consistent Data... 40
Time of Day for Treatments and Exposures 42
The Confounding Effects of Anesthesia... 42
Completely Thaw Frozen Plasma and Re-Suspend for Cleaner Data 43
Section 3: Making Sense of the Data.. 44
Collect the Endocrine and Other Organs for Retrospective Analysis....... 44
The Bimodal Distribution 46
Circulating Tumors Cells Don't Stay on One Side of the Mouse............ 47
The Non-ubiquitous Activation of a Ubiquitous Artificial Promoter 47
Varied Mechanisms Can Yield the Same Phenotype:
A Shrinking Tumor 48
The Gut Microbiome Affects Cancer.. 49
References.. 51

Index.. 55

Author Biography

David H. Nguyen, Ph.D., received the Department of Defense Predoctoral Fellowship in Breast Cancer Research while conducting doctoral research at the University of California, Berkeley, and Lawrence Berkeley National Laboratory. He then did postdoctoral research jointly between New York University and Lawrence Berkeley National Laboratory. He is a lecturer for the Integrated Science Program that is part of Southern California University of Health Sciences. He is also the founding editor-in-chief of Cancer InCytes Magazine. His research interests are the hormonal, immune, and epigenetic mechanisms behind why childhood trauma correlates with an increased risk of cancer during adulthood.

Chapter 1
Macrophages, Extracellular Matrix, and Estrogens in Breast Cancer Risk

Outline

Introduction.. 1
Section 1: Estrus Cycle Regulates Macrophages in the Mammary Gland.................... 2
Section 2: The Estrus Cycle Regulates Macrophages in the Ovaries......................... 3
Section 3: The Effects of Estrogens on Macrophages.. 4
Section 4: Macrophages as Targets of Endocrine Disrupting Chemicals..................... 5
Section 5: Estrogen Effects on Macrophage ROS Production................................... 6
Section 6: Complex Interplay Between Macrophages and the Extracellular Matrix................... 7
Section 7: Testing the Estrus-Regulated Macrophage-Dependent Hypothesis
  of Breast Cancer Risk... 10
References.. 16

Abstract The primary purpose of this chapter is to succinctly present evidence that macrophages are one of the important components within the complex mechanism underlying why lifelong exposure to estrogen is a risk factor for breast cancer. In line with the systems biology theme of this book, this chapter underscores the integrated way in which multiple organs are regulated by macrophages and systemic hormones. This systems biology view reveals that breast cancer risk due to life-long estrogen exposure needs to be understood as having multiple biological mechanisms that are distinct, yet deeply intertwined. This chapter develops what will be referred to as the estrus-regulated, macrophage-dependent theory of breast cancer risk due to life-long exposure to estrogen.

Introduction

The primary purpose of this chapter is to succinctly present evidence that macrophages are one of the important components within the complex mechanism underlying why lifelong exposure to estrogen is a risk factor for breast cancer. In line with the systems biology theme of this book, this chapter underscores the integrated way in which multiple organs are regulated by macrophages and systemic hormones. This systems biology view reveals that breast cancer risk due to life-long estrogen exposure needs to be understood as having multiple biological mechanisms that are

© Springer International Publishing Switzerland 2016   1
D.H. Nguyen, *Systems Biology of Tumor Physiology*, SpringerBriefs
in Cancer Research, DOI 10.1007/978-3-319-25601-6_1

distinct, yet deeply intertwined. This chapter develops what will be referred to as the estrus-regulated, macrophage-dependent theory of breast cancer risk due to life-long exposure to estrogen.

The secondary purpose of this chapter is to present an example of how a systems biology perspective provides an integrated understanding of the ways in which mammalian physiology is intricately tied to tumor development. The advantage of an integrated perspective is that it reveals more mechanisms that can become candidates for therapeutic intervention, since rarely, if ever, is there just one biological mechanism that is responsible for a complex disease phenotype. The systems biology perspective also guides the scope and strategy of how the research literature can be approached. This will be helpful to investigators who are just starting to study physiological mechanisms that pervade the entire organism. It is also helpful to those who study minute molecular/subcellular mechanisms but who seek to have a global understanding of how physiological mechanisms interact to affect their research question.

Epidemiological studies show a correlation between the following: early onset of puberty and breast cancer risk; late onset of menopause and breast cancer risk; exposure to endocrine disrupting chemicals and the early onset of puberty (Collaborative Group on Hormonal Factors in Breast Cancer 2012). The question of interest now becomes about how we can mitigate breast cancer risk by modulating the biological mechanisms behind these risk factors. This chapter attempts to provide the reader with an overview of why macrophages may be a good candidate as targets of prevention and therapeutic modulation. The last section of the chapter has several goals: (1) propose ways of directly testing the theory that is outlined in this chapter via experiments; (2) propose ways of indirectly testing the theory by presenting verifiable predictions of breast biology in different populations of people; (3) discuss the epidemiology studies relating age, breast density, and the theory outlined in this chapter.

Section 1: Estrus Cycle Regulates Macrophages in the Mammary Gland

The focus of this section is to describe how the estrus cycle regulates the involvement of macrophages in the mammary gland. The estrus cycle in mice consists of four stages: proestrus, estrus, metestrus, and diestrus. The number and location of macrophages in the mouse mammary gland fluctuates during these four phases. The total number of macrophages peaks during diestrus, the last stage of the cycle. But with macrophages, location is just as—if not more—important than total number. Macrophages are located nearby and next to the epithelial ducts, but are also found in the adipose regions and connective tissue regions. Epithelial ducts in the mouse mammary gland are like a system of branching tubes that have grape-like clusters (pouches) growing out from the walls of the tubes. The grape-like clusters are called alveolar buds. During the estrus cycle, many epithelial cells in the ducts and buds divide, causing the buds to enlarge. Late in the cycle, many cells in the ducts and buds die, needing to be engulfed by macrophages, and causing the buds to shrink. This growth and shrinkage happens with each cycle. More macrophages congregate around alveolar buds than around the ducts. Proestrus can be considered the preparatory phase between two cycles. Thus, it is the stage when apoptosis occurs in the alveolar buds and the stage that exhibits the highest concentration of macrophages surrounding the epithelial cells (Chua et al. 2010). The same research group that characterized the location and frequency of macrophages in the mammary gland during the estrus cycle also characterized certain phenotypic markers of those macrophages. CD204 is a cell surface receptor on macrophages that helps them recognize dying cells that need to be engulfed. MHCII is a cell surface marker on macrophages that presents engulfed antigens to naïve CD4+ or CD8+ T lymphocytes. The expression of CD204 and MHCII peaked during proestrus and then subsided during metestrus and diestrus (Hodson et al. 2013). These peaks during proestrus are consistent with the increased apoptosis in epithelial cells that occurs during this phase (Chua et al. 2010), meaning that the macrophages are actively engulfing dying cells. NKG2D is a cell surface marker expressed on macrophages that helps them detect dying cells that have DNA damage. In response to these dying cells, NKG2D helps macrophages release proinflammatory cytokines that recruit other leukocytes to the injured area. Opposite to the pattern of CD204 and MHCII expression, macrophages had the highest expression of NKG2D during the last two phases of the estrus cycle, metestrus and diestrus (Hodson et al. 2013). Hodson, Chua, Ingman and colleagues showed that the ovarian hormone progesterone was one of the key regulators responsible for the CD204, MHCII, and NKG2D expression phenotypes (Hodson et al. 2013).

Macrophages play essential roles during mammary gland development. Mouse models in which macrophages were removed exhibited a phenotype in which the mammary glands did not branch into the fat pad during puberty (Gouon-Evans et al. 2000). Not only are macrophages involved in normal ductal development, they are crucial factors in tumor development. Macrophages are required in order for a small mammary tumor to undergo the "angiogenic switch" during which the blood vessel network that feeds the tumor dramatically expands and allows the tumor to enlarge in size (Lin et al. 2006).

In summary, macrophage frequency and function within the mammary gland are systematically regulated during the estrus cycle.

Section 2: The Estrus Cycle Regulates Macrophages in the Ovaries

It is important to note the crucial role of macrophages in normal ovarian function. This section describes how macrophages are essential mediators of normal ovarian function, which in turn is an essential mediator of normal mammary gland function. Thus, the increased risk for breast cancer due to the function and dysfunction of macrophages in the mammary gland is intricately tied to the function and dysfunction of macrophages in the ovaries.

The ovaries contain the egg cells that cyclically mature and ovulate through each estrus cycle. Macrophages are involved in the maturation of the egg follicle, the process of ovulation, and the process of forming the corpus luteum after the ovulation event. The fluctuation of macrophages within the ovary throughout the estrus cycle has been documented (Brannstrom et al. 1993, 1994). Macrophages secrete many products that aid in ovarian function. They secrete serine proteases, cysteine proteases, and matrix metalloproteases. These proteases degrade the basement membrane and extracellular matrix, which is part of the mechanism of how mature follicles rupture during ovulation to release the egg. Macrophages also release growth factors that promote the maturation of the follicle. After ovulation, macrophages transform the remains of the follicular sac into the corpus luteum, which produces the high levels of progesterone during the luteal phase of the estrus cycle. Macrophages then degrade the corpus luteum by engulfing apoptotic cells that appear at the end of the estrus cycle (reviewed in (Wu et al. 2004)).

In summary, macrophages play essential roles in normal ovarian function, meaning a dysregulation of macrophage function in the ovaries would lead to abnormal mammary gland function since the ovarian hormones control the mammary gland. Section 4 of this chapter, entitled "Macrophages as Targets of Endocrine Disrupting Chemicals" applies to both the macrophages in the ovaries and the mammary glands.

Section 3: The Effects of Estrogens on Macrophages

Estrogens have been shown suppress pro-inflammatory activity in macrophages and neutrophils by reducing the production of the cytokines CXCL8, IL-6, TNFα, MIF (Hsieh et al. 2007; Messingham et al. 2001; Pioli et al. 2007; Suzuki et al. 2008). Both ER-α and ER-β are expressed in monocytes and macrophages (Ashcroft et al. 2003; Khan et al. 2005; Kramer and Wray 2002; Mor et al. 2003; Pioli et al. 2007; Wang et al. 2006). Mor and colleagues showed that cultured U937 human monocyte cells express ER-β, but turn off ER-β and switch to ER-α when they differentiate into macrophages after treatment with phorbal ester. ER-α becomes the dominant isoform in differentiated macrophages. Treatment of U937 differentiated macrophages, but not the pre-differentiated monocytes, with estradiol caused apoptosis. Blocking the activity of ER-α with tamoxifen reversed the apoptotic effect of estrogen on macrophages (Mor et al. 2003). Consistent with this induction of apoptosis by estradiol treatment, the number of circulating monocytes increases in women during menopause, while hormone replacement therapy in this group reduced the monocyte count to levels found in younger women (Ben-Hur et al. 1995). The findings of the above Mor et al. study were echoed by a later study from Murphy and colleagues. Murphy et al. showed that freshly isolated human monocytes expressed more ER-β than ER-α, while monocytes differentiated into macrophages after 7 days of GM-CSF treatment exhibited a switch from ER-β to ER-α as the dominant isoform. Furthermore, they showed that macrophages predominantly expressed a truncated ER-α isoform called ERα46 compared to the full-length ERα66 isoform. Monocytes, on the other hand, expressed equivalent amounts of ERα46 and ERα66. Treatment with estradiol induced ER-α levels in macrophages, but not ER-β (Murphy et al. 2009).

Section 4: Macrophages as Targets of Endocrine Disrupting Chemicals

Since macrophages express estrogen receptors and estrogen-related receptors, these proteins are some of the prime candidates for studying the adverse effects of endocrine disrupting chemicals (EDCs). EDCs are chemicals that naturally occur in nature or were synthesized by humans, and that have the ability to mimic or block the activity of natural hormones in the body. Common EDCs are found in plastic food containers, pesticides and herbicides, flame retardants, and industrial chemicals (Schug et al. 2011).

Ohnishi and colleagues examined the effect of 37 EDCs on the ability of macrophages to activate the interferon-beta (IFNβ) promoter after stimulation with lipopolysaccharide (LPS). They tested two classes of EDCs, resin-related chemicals and agrochemicals, on the mouse macrophage cell line RAW264 that harbored an IFN-β-dependent luciferase reporter construct. Among the chemicals were atrazine, bisphenol A (BPA), multiple phthalates, and multiple phenols. The authors reported that most chemicals inhibited the LPS-stimulation of the IFNβ promoter when used within the 50–200 mM range, which the authors note as being above the typical levels found in the environment (Ohnishi et al. 2008). A prior study from the same research group examined the effect of 18 different EDCs on the ability of LPS-stimulated mouse macrophages to produce TNFα and nitric oxide (NO). The results indicate while many EDCs affect macrophage activation by the endotoxin LPS, the nature of the effect is dependent upon the type of EDC and the in vitro (RAW264 cell line) or ex vivo (peritoneal macrophages) context of the experiment (Hong et al. 2004).

Bisphenol A, also known as BPA, is a synthetic estrogen that is commonly used in polycarbonate plastics and the epoxy resins that line food and beverage containers (reviewed in (Vandenberg et al. 2007)). A study of canned tomatoes sold in supermarkets in Italy found that BPA was detected in 22 out of 42 brands (Grumetto et al. 2008). A study of reusable water bottles found that incubating water at room temperature for 5 days caused BPA to leach from the plastic. Furthermore, heating the water contained in steel bottles that were lined with resin increased the amount of leached BPA (Cooper et al. 2011).

A study by Melzer and colleagues found that BPA levels detected in the urine of human males associated with gene expression changes in immune cells. BPA induced the mRNA of ER-β and ERR-α in peripheral blood leukocytes, which includes monocytes (Melzer et al. 2011). Monocytes are the precursor cells that turn into macrophages after they exit the blood stream and reside a tissue. In mouse macrophages, ERR-α modulates the production of reactive oxygen species (ROS) in response to infection (Sonoda et al. 2007). See the topic entitled "Estrogen Effects on Macrophage ROS Production" for a discussion of the role of macrophage-derived ROS in cancer. Relating to the discussion of prolonged estrogen signaling as a risk factor of breast cancer in women, it remains to be determined if ERR-α is induced by BPA in the peripheral blood leukocytes of women. The study by Melzer and colleagues (2011), which looked at transcript levels in lymphocytes from men, found a positive correlation between the co-induction of ER-β and ERR-α by BPA. This suggests that BPA induces a coordinated module of genes that includes at least ER-β and ERR-α, which may also operate similarly in women.

It is important to note that BPA and other EDCs affect immune cell types other than macrophages, such as lymphocytes, and that these other types of cells interact with and influence macrophage phenotypes. Furthermore, in addition to the estrogen and estrogen-related receptor families, BPA also interacts with the androgen, aryl hydrocarbon, and peroxisome proliferator-activated receptor families (reviewed in (Rogers et al. 2013)).

Section 5: Estrogen Effects on Macrophage ROS Production

Reactive oxygen species (ROS) are unstable molecules containing an oxygen atom that has an unpaired electron that disrupts covalent bonds on other molecules. This disruption by ROS causes DNA, RNA, lipid, and protein damage, which can contribute to genomic instability of a cell. Macrophages normally produce ROS and reactive nitrogen species (RNS) during the process of fighting infection. After engulfing bacteria, macrophages digest them by fusing the vesicle that contains the bacteria with vesicles that contain digestive enzymes, lysozymes, or ROS (reviewed in (Horta et al. 2012; Slauch 2011)). However, when macrophages persist in a tissue longer than is needed, they contribute to the problem of chronic inflammation. In chronic inflammation, macrophages and other immune cells remain in a tissue and continually produce low to moderate levels of ROS and cytokines. The persistent presence of ROS around cells of a normal tissue can eventually lead to DNA damage within the normal tissue. Furthermore, the cytokines produced by macrophages bind to normal tissue cells and activate intracellular signaling pathways. These pathways can lead to the induction of genes that degrade DNA, and the suppression of genes that repair mismatched DNA. The result is that normal cells in a chronic inflammation environment accumulate more replication errors than their counterparts in a normal environment (reviewed in (Elinav et al. 2013; Liou and Storz 2010; Waris and Ahsan 2006).

Estrogens generally have a suppressive effect on macrophages and other immune cell types. They promote apoptosis in monocytes and macrophages (reviewed in (Roy et al. 2007)). A number of studies indicated that estrogens significantly influence the production of ROS in macrophages. 17beta-estradiol induced calcium influx into RAW-264.7 murine macrophage cells, which coincided with an increased production of ROS (Azenabor and Chaudhry 2003). Peritoneal macrophages from male rats were induced to produce ROS when treated with estrogen, but only at concentrations below or above the range between 10^{-10} to 10^{-9} M (Chao et al. 1994), suggesting a multi-modal response to estrogen. Two studies from the same research group showed that removing the ovaries from female mice reduced the activity of antioxidant enzymes in macrophages. Furthermore, this reduced activity could be restored by treatment with exogenous estrogen. The first study showed that estrogen levels directly correlated with macrophage production of hydrogen peroxide (Azevedo et al. 1997) while the second showed that estrogen levels directly correlated with catalase activity in macrophages (Azevedo et al. 2001).

In summary, there is ample evidence suggesting that the surges of estrogen during estrus cycling regulates both the factors that produce and quench ROS, which is a key feature of the chronic inflammation that promotes cancer. It should be noted that Sect. 6 of this chapter discusses the effect of a stiff physical environment on macrophages, and that stiffness has been reported to increase ROS production (Patel et al. 2012) meaning that a deregulation of estrogen signaling may compound the effects of an abnormally stiff microenvironment on macrophage function. As a side note, it is interesting to note that ovariectomy was reported to result in more peritoneal macrophages with chromosomal aberrations, which was reversed with estrogen treatment (Lacava and Luna 1994). It would not be a stretch to hypothesize that mutated macrophages would be more prone to incorrect estrogen-regulated ROS activity. In light of the section in this chapter about the effect of endocrine disrupting chemicals, one would predict that circulating monocytes or in situ macrophages from humans who were exposed to higher levels of inhibitory estrogenic endocrine disrupting agents would harbor more—or a distinct profile of—chromosomal aberrations.

Section 6: Complex Interplay Between Macrophages and the Extracellular Matrix

Composition of the Extracellular Matrix

The extracellular matrix (ECM) is a network of protein and carbohydrate molecules that form the scaffold in which cells live and move. It is analogous to the wooden frame within the walls of a house or the highways, walls, and bridges within a city. The ECM affects the cells' decisions to move, settle down, divide, or die. This matrix is composed of different types of collagen molecules, fibronectin, elastin, decorin, aggrecan, and perlecan (Frantz et al. 2010). In addition to all the cell types within the mammary gland, macrophages also communicate with the ECM. As the physical properties (i.e. stiffness, elasticity, etc.) of the ECM change because of aging, damage, and disease, the messages that it sends to macrophages and other cell types begin to change. These altered physical signals dramatically affect the behavior of cells.

Macrophages Modify the Mammary Extracellular Matrix

Macrophages not only reside within the ECM of the mammary gland, but they play important roles in digesting and rearranging the ECM. During puberty in the mouse, the mammary epithelial tree cannot grow to fill the entire gland if macrophages are removed (Gouon-Evans et al. 2000). Puberty is the time during which the small ductal tree that is already present in the mammary gland begins to extend and branch, resulting in a tree-like structure as the mouse enters adulthood. At the leading end of each arm in this expanding tree is a large bulb of cells called the tail end bud (TEB). The TEB is what pushes through the fat tissue in the gland, is where new cells are formed that will line the growing ducts, and is where the arm of the tree occasionally splits to form two new branches. Macrophages congregate at the base, or the neck, of the TEB. Ingman and colleagues conducted a study that examined the interaction of macrophages around the TEB with the ECM that surrounds the TEB. The study revealed that the developing mammary gland has long collagen fibers that run along its entire length. These fibers are also present around the neck of the TEBs, radiating from the neck towards the direction of the TEB's growth. Younger TEBs also have collagen fibers that radiate perpendicularly from the neck region. It turns out that macrophages help organize collagen molecules into these large fibers. Macrophages move along these fibers as if on train tracks, but do also move across them. When these large fibers are present, macrophages move faster within the mammary gland (Ingman et al. 2006).

Post-partum involution is another period of normal mammary gland development in which macrophages mediate a remodeling of the ECM. After a mouse has given birth to pups and has stopped breastfeeding them, her mammary glands undergo a massive reduction in size and content. Like in humans, the mouse mammary gland enlarges and fills with milk in preparation for the offspring. The enlargement results from cells in the epithelial tree dividing to create more tunnel space within the ductal system. This space then swells full with milk. When the pups are grown and breastfeeding (lactation) is no longer needed, many of the cells in the ductal system die. But cell death alone does not shrink the mammary gland back to normal. The ECM that accompanied, and was distinct to, the enlarged gland during pregnancy needs to be remodeled back to a pre-pregnancy state. O'Brien and colleagues conducted a study in which they examined the phenotype of macrophages before, during, and after involution. During involution, the mammary gland contained an eightfold increase in the number of macrophages compared to nulliparous female mice. Furthermore, these macrophages exhibited a pro-tumor phenotype similar to those involved in wound healing and the promotion of angiogenesis. Interestingly, cell culture experiments showed that ECM extracted from involuting mammary glands attracted macrophages much more than ECM from glands of nulliparous mice. The authors also reported that in cell culture experiments, denatured collagen 1 was also chemotactic, attracting macrophages to migrate towards the source of the collagen. In all, the study presented evidence that macrophages are highly recruited to the mammary gland during involution (O'Brien et al. 2010). Within the context of macrophages that exhibit the wound healing phenotype, these macrophages are known to play key roles in digesting the collagen in the ECM of wounds (Madsen et al. 2013; Madsen and Bugge 2013).

Effect of Increased Matrix Stiffness on Breast Cancer Progression

Before the discussion of how altered stiffness of the ECM affects macrophage phenotypes that then affect breast cancer cells, it is helpful to understand the effect of ECM stiffness on the cancer itself. The subfield of breast cancer that studies how the physical environment around a cancer cell, that is the "microenvironment," can subdue that cancer was pioneered by the laboratory of Mina Bissell, Ph.D. Extending this idea into a subfield that studies the mechanical properties of the breast cancer microenvironment is the pioneering work of the laboratory of Valerie Weaver, Ph.D. Two studies from the Weaver laboratory will suffice to illustrate the importance of mechanical forces on breast cancer cells. Levental and colleagues performed elegant experiments showing that as the surrounding tissue became stiffer, the more the breast cancer developed (Levental et al. 2009). Using the MMTV-Neu mouse model of breast cancer, which has normal mammary ducts that progress to small tumors and then to large tumors, the authors showed that the ECM increasingly stiffens and the collagen in the ECM increasingly linearizes as the pre-cancerous lesions develop into advanced tumors. They also showed that lysyl oxidase (LOX), an enzyme that crosslinks collagen molecules in the ECM, was a key feature of this enhanced tissue stiffness. Forcing the expression of LOX in fibroblasts that were then co-injected with pre-malignant breast cells (MCF10AT) into mice resulted in stiffer tumors compared to those that arose from the same type of breast cells that were co-injected with un-modified fibroblasts. Furthermore, the authors showed that tumor growth could be inhibited by administering either a chemical that blocked LOX activity (BAPN) or an anti-LOX antibody. The group that produced this paper also conducted another study in which they revealed a link between the stiffness of the microenvironment and reduced levels of the tumor suppressor called phosphatase and tensin homolog (PTEN) (Mouw et al. 2014). Mouw and colleagues deciphered that by increasing the stiffness of the environment, they could induce the expression of a micro-RNA called miR-18a that targeted PTEN. MiR-18a reduced levels of PTEN in two ways. First, it directly bound to PTEN mRNA and inhibited PTEN translation. Secondly, miR-18a reduced levels of the transcription factor homeobox A9 (HOXA9), which transcriptionally activated PTEN. These two studies provide many insights into how normal and malignant breast cells sense physical forces in their microenvironment, and how the stiffness of the physical environment promotes cancer progression.

Effect of Increased Matrix Stiffness on Macrophage Phenotypes

Macrophages not only remodel the ECM, but their behaviors are dramatically affected by the mechanical properties of the ECM in which they reside. Wehner and colleagues cultured peritoneal macrophages on a platform that could be mechanically stretched (Wehner et al. 2010). When they applied 6 h of stretching force to the platform and thus the macrophages that were attached to that platform, they activated an inflammatory program that was not present in the non-stretched control group. The authors also determined that stretching the macrophages while exposing them to the bacterial endotoxin LPS synergistically induced several inflammation genes two- to three-fold higher than LPS exposure alone: inducible nitric oxide synthase (iNOS), cyclooxygenase (COX-2), and interleukin-6 (IL-6). These data are in accord with a study done by Patel and colleagues, which showed that culturing mouse or human alveolar macrophages on stiff compared to soft substrates altered their behaviors (Patel et al. 2012). The authors cultured the macrophages on acrylamide gels that had the softness of lung tissue (1.2 kPa) or the stiffness of fibrotic tissue (150 kPa). Culturing on stiff gels caused the macrophages to exhibit increased phagocytosis of beads or bacteria, and also increased production of ROS. Blakney and colleagues also explored the combined effects of substrate stiffness and infection on macrophage phenotypes (Blakney et al. 2012). Using Poly(ethylene glycol) (PEG) hydrogels that varied in stiffness from 130, 240, or 840 kPa, and that were implanted subcutaneously in mice, the authors found that the stiffer gels had more macrophages attached to their surface 28 days after implantation. In vitro, after stimulation with LPS, macrophages that were attached to stiffer gels had a higher expression of tumor necrosis factor-alpha (TNF-α), interleukin-1-beta (IL-1β), and IL-6 compared to those that were attached to softer gels. Thus, there is a consistent theme that a stiffer microenvironment enhances the degree by which macrophages are activated in response to insults.

The above studies indicate that macrophage phenotype is determined by a combination of the physical signals that come from the microenvironment and other cues such as infection. Thus, it will be critical for the field of breast cancer research to understand the effect of the physical microenvironment on the behavior of macrophages. Macrophage phenotypes are complex and exhibit plasticity, meaning they can transition between the different general phenotypes, i.e. would healing, anti-microbial, and anti-inflammatory (reviewed in (Mosser and Edwards 2008)). Further studies elucidating the effects of the physical environment on macrophage function are certainly warranted. As more mechanisms of action within more subsets of macrophages defined by their physical environment are revealed, general themes will emerge that will guide the development of new preventive therapeutics.

Section 7: Testing the Estrus-Regulated Macrophage-Dependent Hypothesis of Breast Cancer Risk

Directly and Indirectly Testing the Hypothesis

Up to this point, the content of this chapter was meant to lay the foundation for why macrophages should be considered crucial preventive and therapeutic targets in the biological mechanisms underlying the risk of breast cancer from life-long exposure to

menopause, or gradually develop density up through menopause after which the density level is maintained or lost. As with many aspects of physiology, there are likely subgroups of women with regard to this phenotype. Radiographic breast density, as measured by the Checka study and others, is defined as the ratio of the area represented by epithelial and stromal cells, which appears white, compared to fatty tissue, which appears radiolucent (varying shades of gray). The Checka study involved patients that were mostly Caucasian and who had medical insurance. It is well known that racial disparities significantly affect breast density, along with the age of onset of breast cancer and the type of breast cancer that develops. African-American women have higher rates of triple-negative breast cancers (Kurian et al. 2010), the most aggressive type of breast cancer. Thus, race is likely another important factor in the rate of density gain and loss.

Breast tissue from reduction mammoplasty or other surgical procedures from women of varying ages would be great for examining the structural arrangement of collagen in the extracellular matrix. An increase of fibrillar collagen and stiffness with increasing age would support the theory outlined in this chapter.

The Effect of Premature Ovarian Failure

A second question for indirectly testing the theory is to ask whether women with premature ovarian failure have less dense breast tissue than normal women, since those who live beyond the ovarian failure would have had fewer menstrual cycles than normal women of the same age. Benetti-Pinto and colleagues (2014) examined the mammographic density of 56 women who experienced premature ovarian failure (POF) at an average age of 32.35 years. Each subject had two mammograms that were taken an average of 5.25 years apart. Consistent with the theory outlined in this chapter, the authors found that the breast density of women with POF decreased across 5 years, regardless of whether they used estrogen-progestin therapy. The same research group conducted a similar study 6 years prior to this one in which they did a one-time-point comparison of mammographic density between POF patients who were on hormone therapy and normal women (Benetti-Pinto et al. 2008). In that study they did not find a difference in breast density between the POF group and the normal controls, suggesting that POF does not decrease breast density. However, the findings from the 2014 study nullify the results of the 2008 study, because the 2014 study revealed that parity significantly reduces the degree of breast density in POF patients. In the later 2014 study 32 % (18/56) of POF patients were nulliparous (never having had a full-term pregnancy) while those who were multiparous (having had more than one full-term pregnancy) were shown to have a lesser degree of breast density. In other words, having had children decreases women's degrees of breast density. This is significant for the comparison of the two studies because the earlier 2008 study reported 54.8 % of the POF patients as "breastfeeding," meaning that they had at least one full-term pregnancy. The fact that half of the POF group in the 2008 study was parous means that the POF cohorts between the two studies are not comparable. Further evidence for this can be found in the two papers. In the 2008 study, the POF group had an average age of 36.9 years and an average digitized breast density of 25.1 %. In the 2014 study, the nulliparous POF group at the second mammography screening had an average age of 40.33 years and an average digitized breast density of 25.4 %. It is known that mammographic density decreases with age (Checka et al. 2012), so even though the POF group in the 2008 study was several years younger, it had a comparable degree of breast density as the POF group in the 2014 study; likely due to the fact that half of the 2008-POF group was parous, which correlates with decreased breast density. The effect of parous POF patients also applies to the third study from this research group, which was published in 2010 (Soares et al. 2010). Soares and colleagues compared the mammographic density of POF patients who were on hormone therapy to normal post-menopausal women who were on hormone therapy, finding no difference. However, as in their 2008 study, half of the POF cohort had breastfed and were thus parous. Additionally, in the Soares study the patients in the POF group had already been diagnosed for an average of 85.9 months (7.16 years) at the time of the study. 7.16 years may have been a time during which significant reduction in breast density in the POF cohort had already occurred, since the 2014 study showed this effect. In summary, further studies with larger sample sizes and comparable experimental design are warranted to clarify the effect of POF on mammographic density.

As with the data linking aging to breast density, mammography does not reveal the structural characteristics of the extracellular matrix nor the physical properties of the tissue. Measuring these factors in breast tissue from women several years after they underwent POF compared to age-matched normal women would be a way of confirming the predictions of the theory outlined in this chapter.

The Effect of Hormone Replacement Therapy in Post-menopausal Women

A third question for indirectly testing the theory outlined in this chapter is to ask if hormone replacement therapy in post-menopausal women increases breast density. Exogenously treating a patient with estrogens and progestins is akin to an "overexpression" experiment in which the hypothesized effect is induced by artificially adding or promoting a factor. In this case, the theory outlined in this chapter would predict that treatment with exogenous hormones would increase breast density in post-menopausal women. Indeed, multiple studies have shown this to be true, have reported that varying doses of hormone treatment correlate with varying degrees of increased density, and that not all women who receive treatment exhibit the same degree of increased density (Christodoulakos et al. 2006; Laya et al. 1995; Marugg et al. 1997). Though the post-menopausal patients in these studies exhibit increased mammographic density due to hormone therapy, it is unknown if macrophages are recruited to their mammary tissue as part of why the density increased. An increase in the amount of epithelial cells, stromal cells, and extracellular matrix would all result in increased mammographic density. However, it is unknown whether the new extracellular matrix in these post-menopausal women is arranged in ways that are similar to the effect of macrophages on the matrix during normal menstruation.

The Effect of Wiskott–Aldrich Syndrome on Breast Density

A fourth question for indirectly testing the theory is to ask if women with mutated macrophages have less dense breasts than women with normal macrophages. The Wiskott–Aldrich Syndrome (WAS) is an X-linked disorder that results in people who have a mutated form of the Wiskott–Aldrich Syndrome protein (WASP). The WASP protein is involved in actin polymerization, which is critical for cell movement. Since the WASP gene is primarily expressed in hematopoietic cells, WAS patients exhibit recurring infections, easy bruising, prolonged bleeding, and a low number of small platelets (reviewed in (Thrasher and Burns 2010)). Since macrophages are derived from the hematopoietic system, WAS patients should exhibit deficits or impairments regarding physiological features that are governed my macrophages. Macrophages from patients with WASP mutations exhibit impaired filopodia, lamellapodia, and cell migration (reviewed in (Thrasher and Burns 2010)). Given the role of macrophage in remodeling the ECM of the mammary gland, the theory outlined in this chapter would predict that WAS patients would have breast tissue that is less dense and that has altered extracellular matrix, compared to normal women, because of impaired macrophage function. The advent of modern medical treatments such as bone marrow transplants and antibiotics has allowed WAS patients to survive for multiple decades, which is enough time for their breast tissue to have undergone many menstrual cycles (Blaese et al. 2013). Since WAS patients may have little concern for developing breast cancer, which is a disease significantly influenced by age, there would be shortage of mammography data with which to examine breast density. However, due to the types of cancers that WAS patients experience and die from (Salavoura et al. 2008), they may commonly undergo thoracic computed tomography (CT) scans for diagnostic purposes. The CT data would allow investigators to determine the density of their breast tissue as a function of age. CT scans are relatively comparable to mammography for determining breast density (Salvatore et al. 2014). Retrospective analysis of breast density via CT data should consider and extrapolate the menstrual phase of the patient at the time of the scan. WAS patients who undergo bone marrow transplantations may have most of their myeloid-derived cells, which includes macrophages, replaced, which may be a confounding factor in determining whether the WAS affects breast density based on dysfunction macrophages. However, differing WAS patients will have received bone marrow transplants at different times and thus may still have had significant years of adult life with dysfunctional macrophages affecting the mammary glands. Furthermore, not all bone marrow graft procedures yield a completely recovered immune system after the same amount of time, which may still leave a graft recipient with a number of years with dysfunctional mammary macrophages.

Closing Thoughts

In testing my hypothesis, I propose the use of imaging modalities that capture gross features of breast tissue. These modalities cannot pinpoint microscopic mammary epithelial lesions that may be increasing in stiffness and matrix reorganization even though the density of the overall breast may be decreasing as the woman ages. Perhaps new imaging modalities in the future will allow us to detect individual foci of stiffness and matrix reorganization that worsen over the lifetime of the patient. Also, there is evidence that in utero exposure to estrogen-mimicking compounds can cause epigenetic marks that then increase the risk of breast cancer after the woman is born. The details of those studies are not discussed here, but suffice it to say that an in utero epigenetic theory of breast cancer risk across a person's lifetime goes hand-in-hand with the estrogen-regulated, macrophage-dependent theory of breast cancer risk that is outlined in this chapter. Microscopic lesions that develop over a woman's lifetime from epigenetically altered mammary epithelial cells will inevitably alter the macrophages that they encounter, along with the adjacent extracellular matrix.

References

Ashcroft, G. S., Mills, S. J., Lei, K., Gibbons, L., Jeong, M. J., Taniguchi, M., et al. (2003). Estrogen modulates cutaneous wound healing by downregulating macrophage migration inhibitory factor. *The Journal of Clinical Investigation, 111*, 1309–1318.

Azenabor, A. A., & Chaudhry, A. U. (2003). 17 beta-estradiol induces L-type Ca2+ channel activation and regulates redox function in macrophages. *Journal of Reproductive Immunology, 59*, 17–28.

Azevedo, R. B., Lacava, Z. G., Miyasaka, C. K., Chaves, S. B., & Curi, R. (2001). Regulation of antioxidant enzyme activities in male and female rat macrophages by sex steroids. *Brazilian Journal of Medical and Biological Research, 34*, 683–687.

Azevedo, R. B., Rosa, L. F., Lacava, Z. G., & Curi, R. (1997). Gonadectomy impairs lymphocyte proliferation and macrophage function in male and female rats. Correlation with key enzyme activities of glucose and glutamine metabolism. *Cell Biochemistry and Function, 15*, 293–298.

Benetti-Pinto, C. L., Brancalion, M. F., Assis, L. H., Tinois, E., Giraldo, H. P., Cabello, C., et al. (2014). Mammographic breast density in women with premature ovarian failure: A prospective analysis. *Menopause, 21*, 933–937.

Benetti-Pinto, C. L., Soares, P. M., Magna, L. A., Petta, C. A., & Dos Santos, C. C. (2008). Breast density in women with premature ovarian failure using hormone therapy. *Gynecological Endocrinology, 24*, 40–43.

Ben-Hur, H., Mor, G., Insler, V., Blickstein, I., Amir-Zaltsman, Y., Sharp, A., et al. (1995). Menopause is associated with a significant increase in blood monocyte number and a relative decrease in the expression of estrogen receptors in human peripheral monocytes. *American Journal of Reproductive Immunology, 34*, 363–369.

Bertrand, K. A., Tamimi, R. M., Scott, C. G., Jensen, M. R., Pankratz, V., Visscher, D., et al. (2013). Mammographic density and risk of breast cancer by age and tumor characteristics. *Breast Cancer Research, 15*, R104.

Blaese, M. R., Bonilla, F. A., Stiehm, E. R., & Younger, E. M. (Eds.). (2013). *Wiskott–Aldrich syndrome* (Patient & family handbook for primary immunodeficency diseases, pp. 54–60). Towson, MD: Immune Deficiency Foundation.

Blakney, A. K., Swartzlander, M. D., & Bryant, S. J. (2012). The effects of substrate stiffness on the in vitro activation of macrophages and in vivo host response to poly(ethylene glycol)-based hydrogels. *Journal of Biomedical Materials Research, Part A, 100*, 1375–1386.

Boyd, N. F., Guo, H., Martin, L. J., Sun, L., Stone, J., Fishell, E., et al. (2007). Mammographic density and the risk and detection of breast cancer. *The New England Journal of Medicine, 356*, 227–236.

Brannstrom, M., Mayrhofer, G., & Robertson, S. A. (1993). Localization of leukocyte subsets in the rat ovary during the periovulatory period. *Biology of Reproduction, 48*, 277–286.

Brannstrom, M., Pascoe, V., Norman, R. J., & McClure, N. (1994). Localization of leukocyte subsets in the follicle wall and in the corpus luteum throughout the human menstrual cycle. *Fertility and Sterility, 61*, 488–495.

Byng, J. W., Yaffe, M. J., Jong, R. A., Shumak, R. S., Lockwood, G. A., Tritchler, D. L., et al. (1998). Analysis of mammographic density and breast cancer risk from digitized mammograms. *Radiographics, 18*, 1587–1598.

Chao, T. C., Van Alten, P. J., & Walter, R. J. (1994). Steroid sex hormones and macrophage function: Modulation of reactive oxygen intermediates and nitrite release. *American Journal of Reproductive Immunology, 32*, 43–52.

Checka, C. M., Chun, J. E., Schnabel, F. R., Lee, J., & Toth, H. (2012). The relationship of mammographic density and age: Implications for breast cancer screening. *AJR. American Journal of Roentgenology, 198*, W292–W295.

Christodoulakos, G. E., Lambrinoudaki, I. V., Vourtsi, A. D., Vlachou, S., Creatsa, M., Panoulis, K. P., et al. (2006). The effect of low dose hormone therapy on mammographic breast density. *Maturitas, 54*, 78–85.

Chua, A. C., Hodson, L. J., Moldenhauer, L. M., Robertson, S. A., & Ingman, W. V. (2010). Dual roles for macrophages in ovarian cycle-associated development and remodelling of the mammary gland epithelium. *Development, 137*, 4229–4238.

Collaborative Group on Hormonal Factors in Breast Cancer. (2012). Menarche, menopause, and breast cancer risk: Individual participant meta-analysis, including 118 964 women with breast cancer from 117 epidemiological studies. *The Lancet Oncology, 13*, 1141–1151.

Cooper, J. E., Kendig, E. L., & Belcher, S. M. (2011). Assessment of bisphenol A released from reusable plastic, aluminium and stainless steel water bottles. *Chemosphere, 85*, 943–947.

Elinav, E., Nowarski, R., Thaiss, C. A., Hu, B., Jin, C., & Flavell, R. A. (2013). Inflammation-induced cancer: Crosstalk between tumours, immune cells and microorganisms. *Nature Reviews Cancer, 13*, 759–771.

Frantz, C., Stewart, K. M., & Weaver, V. M. (2010). The extracellular matrix at a glance. *Journal of Cell Science, 123*, 4195–4200.

Gouon-Evans, V., Rothenberg, M. E., & Pollard, J. W. (2000). Postnatal mammary gland development requires macrophages and eosinophils. *Development, 127*, 2269–2282.

Grumetto, L., Montesano, D., Seccia, S., Albrizio, S., & Barbato, F. (2008). Determination of bisphenol a and bisphenol B residues in canned peeled tomatoes by reversed-phase liquid chromatography. *Journal of Agricultural and Food Chemistry, 56*, 10633–10637.

Hodson, L. J., Chua, A. C., Evdokiou, A., Robertson, S. A., & Ingman, W. V. (2013). Macrophage phenotype in the mammary gland fluctuates over the course of the estrous cycle and is regulated by ovarian steroid hormones. *Biology of Reproduction, 89*, 65.

Hong, C. C., Shimomura-Shimizu, M., Muroi, M., & Tanamoto, K. (2004). Effect of endocrine disrupting chemicals on lipopolysaccharide-induced tumor necrosis factor-alpha and nitric oxide production by mouse macrophages. *Biological & Pharmaceutical Bulletin, 27*, 1136–1139.

Hong, H., Yen, H. Y., Brockmeyer, A., Liu, Y., Chodankar, R., Pike, M. C., et al. (2010). Changes in the mouse estrus cycle in response to BRCA1 inactivation suggest a potential link between risk factors for familial and sporadic ovarian cancer. *Cancer Research, 70*, 221–228.

Horta, M. F., Mendes, B. P., Roma, E. H., Noronha, F. S., Macedo, J. P., Oliveira, L. S., et al. (2012). Reactive oxygen species and nitric oxide in cutaneous leishmaniasis. *Journal of Parasitology Research, 2012*, 203818.

Hsieh, Y. C., Frink, M., Hsieh, C. H., Choudhry, M. A., Schwacha, M. G., Bland, K. I., et al. (2007). Downregulation of migration inhibitory factor is critical for estrogen-mediated attenuation of lung tissue damage following trauma-hemorrhage. *American Journal of Physiology. Lung Cellular and Molecular Physiology, 292*, L1227–L1232.

Ingman, W. V., Wyckoff, J., Gouon-Evans, V., Condeelis, J., & Pollard, J. W. (2006). Macrophages promote collagen fibrillogenesis around terminal end buds of the developing mammary gland. *Developmental Dynamics, 235*, 3222–3229.

Khan, K. N., Masuzaki, H., Fujishita, A., Kitajima, M., Sekine, I., Matsuyama, T., et al. (2005). Estrogen and progesterone receptor expression in macrophages and regulation of hepatocyte growth factor by ovarian steroids in women with endometriosis. *Human Reproduction, 20*, 2004–2013.

Kramer, P. R., & Wray, S. (2002). 17-Beta-estradiol regulates expression of genes that function in macrophage activation and cholesterol homeostasis. *The Journal of Steroid Biochemistry and Molecular Biology, 81*, 203–216.

Kurian, A. W., Fish, K., Shema, S. J., & Clarke, C. A. (2010). Lifetime risks of specific breast cancer subtypes among women in four racial/ethnic groups. *Breast Cancer Research, 12*, R99.

Lacava, Z. G., & Luna, H. (1994). The anticlastogenic effect of tocopherol in peritoneal macrophages of benznidazole-treated and ovariectomized mice. *Mutation Research, 305*, 145–150.

Laya, M. B., Gallagher, J. C., Schreiman, J. S., Larson, E. B., Watson, P., & Weinstein, L. (1995). Effect of postmenopausal hormonal replacement therapy on mammographic density and parenchymal pattern. *Radiology, 196*, 433–437.

Levental, K. R., Yu, H., Kass, L., Lakins, J. N., Egeblad, M., Erler, J. T., et al. (2009). Matrix crosslinking forces tumor progression by enhancing integrin signaling. *Cell, 139*, 891–906.

Lin, E. Y., Li, J.-F., Gnatovskiy, L., Deng, Y., Zhu, L., Grzesik, D. A., et al. (2006). Macrophages regulate the angiogenic switch in a mouse model of breast cancer. *Cancer Research, 66*, 11238–11246.

Liou, G. Y., & Storz, P. (2010). Reactive oxygen species in cancer. *Free Radical Research, 44*, 479–496.

Madsen, D. H., & Bugge, T. H. (2013). Imaging collagen degradation in vivo highlights a key role for M2-polarized macrophages in extracellular matrix degradation. *Oncoimmunology, 2*, e27127.

Madsen, D. H., Leonard, D., Masedunskas, A., Moyer, A., Jurgensen, H. J., Peters, D. E., et al. (2013). M2-like macrophages are responsible for collagen degradation through a mannose receptor-mediated pathway. *The Journal of Cell Biology, 202*, 951–966.

Marugg, R. C., van der Mooren, M. J., Hendriks, J. H., Rolland, R., & Ruijs, S. H. (1997). Mammographic changes in postmenopausal women on hormonal replacement therapy. *European Radiology, 7*, 749–755.

Melzer, D., Harries, L., Cipelli, R., Henley, W., Money, C., McCormack, P., et al. (2011). Bisphenol A exposure is associated with in vivo estrogenic gene expression in adults. *Environmental Health Perspectives, 119*, 1788–1793.

Messingham, K. A., Heinrich, S. A., & Kovacs, E. J. (2001). Estrogen restores cellular immunity in injured male mice via suppression of interleukin-6 production. *Journal of Leukocyte Biology, 70*, 887–895.

Militzer, K., & Schwalenstocker, H. (1996). Postnatal and postpartal morphology of the mammary gland in nude mice. *Journal of Experimental Animal Science, 38*, 1–12.

Mor, G., Sapi, E., Abrahams, V. M., Rutherford, T., Song, J., Hao, X. Y., et al. (2003). Interaction of the estrogen receptors with the Fas ligand promoter in human monocytes. *The Journal of Immunology, 170*, 114–122.

Mosser, D. M., & Edwards, J. P. (2008). Exploring the full spectrum of macrophage activation. *Nature Reviews Immunology, 8*, 958–969.

Mouw, J. K., Yui, Y., Damiano, L., Bainer, R. O., Lakins, J. N., Acerbi, I., et al. (2014). Tissue mechanics modulate microRNA-dependent PTEN expression to regulate malignant progression. *Nature Medicine, 20*, 360–367.

Murphy, A. J., Guyre, P. M., Wira, C. R., & Pioli, P. A. (2009). Estradiol regulates expression of estrogen receptor ERalpha46 in human macrophages. *PLoS One, 4*, e5539.

Nelson, J. F., Felicio, L. S., Randall, P. K., Sims, C., & Finch, C. E. (1982). A longitudinal study of estrous cyclicity in aging C57BL/6J mice: I. Cycle frequency, length and vaginal cytology. *Biology of Reproduction, 27*, 327–339.

O'Brien, J., Lyons, T., Monks, J., Lucia, M. S., Wilson, R. S., Hines, L., et al. (2010). Alternatively activated macrophages and collagen remodeling characterize the postpartum involuting mammary gland across species. *The American Journal of Pathology, 176*, 1241–1255.

Ohnishi, T., Yoshida, T., Igarashi, A., Muroi, M., & Tanamoto, K. (2008). Effects of possible endocrine disruptors on MyD88-independent TLR4 signaling. *FEMS Immunology and Medical Microbiology, 52*, 293–295.

Patel, N. R., Bole, M., Chen, C., Hardin, C. C., Kho, A. T., Mih, J., et al. (2012). Cell elasticity determines macrophage function. *PLoS One, 7*, e41024.

Pinsky, R. W., & Helvie, M. A. (2010). Mammographic breast density: Effect on imaging and breast cancer risk. *Journal of the National Comprehensive Cancer Network, 8*, 1157–1164. quiz 1165.

Pioli, P. A., Jensen, A. L., Weaver, L. K., Amiel, E., Shen, Z., Shen, L., et al. (2007). Estradiol attenuates lipopolysaccharide-induced CXC chemokine ligand 8 production by human peripheral blood monocytes. *The Journal of Immunology, 179*, 6284–6290.

Rogers, J. A., Metz, L., & Yong, V. W. (2013). Review: Endocrine disrupting chemicals and immune responses: A focus on bisphenol-A and its potential mechanisms. *Molecular Immunology, 53*, 421–430.

Roy, D., Cai, Q., Felty, Q., & Narayan, S. (2007). Estrogen-induced generation of reactive oxygen and nitrogen species, gene damage, and estrogen-dependent cancers. *Journal of Toxicology and Environmental Health, Part B: Critical Reviews, 10*, 235–257.

Sabeh, F., Shimizu-Hirota, R., & Weiss, S. J. (2009). Protease-dependent versus -independent cancer cell invasion programs: Three-dimensional amoeboid movement revisited. *The Journal of Cell Biology, 185*, 11–19.

Salavoura, K., Kolialexi, A., Tsangaris, G., & Mavrou, A. (2008). Development of cancer in patients with primary immunodeficiencies. *Anticancer Research, 28*, 1263–1269.

Salvatore, M., Margolies, L., Kale, M., Wisnivesky, J., Kotkin, S., Henschke, C. I., et al. (2014). Breast density: Comparison of chest CT with mammography. *Radiology, 270*, 67–73.

Schug, T. T., Janesick, A., Blumberg, B., & Heindel, J. J. (2011). Endocrine disrupting chemicals and disease susceptibility. *The Journal of Steroid Biochemistry and Molecular Biology, 127*, 204–215.

Sheffield, L. G., & Welsch, C. W. (1988). Transplantation of human breast epithelia to mammary-gland-free fat-pads of athymic nude mice: Influence of mammotrophic hormones on growth of breast epithelia. *International Journal of Cancer, 41*, 713–719.

Slauch, J. M. (2011). How does the oxidative burst of macrophages kill bacteria? Still an open question. *Molecular Microbiology, 80*, 580–583.

Soares, P. M., Cabello, C., Magna, L. A., Tinois, E., & Benetti-Pinto, C. L. (2010). Breast density in women with premature ovarian failure or postmenopausal women using hormone therapy: Analytical cross-sectional study. *São Paulo Medical Journal, 128*, 211–214.

Sonoda, J., Laganiere, J., Mehl, I. R., Barish, G. D., Chong, L. W., Li, X., et al. (2007). Nuclear receptor ERR alpha and coactivator PGC-1 beta are effectors of IFN-gamma-induced host defense. *Genes and Development, 21*, 1909–1920.

Suzuki, T., Yu, H. P., Hsieh, Y. C., Choudhry, M. A., Bland, K. I., & Chaudry, I. H. (2008). Estrogen-mediated activation of non-genomic pathway improves macrophages cytokine production following trauma-hemorrhage. *Journal of Cellular Physiology, 214*, 662–672.

Thrasher, A. J., & Burns, S. O. (2010). WASP: A key immunological multitasker. *Nature Reviews Immunology, 10*, 182–192.

Vandenberg, L. N., Hauser, R., Marcus, M., Olea, N., & Welshons, W. V. (2007). Human exposure to bisphenol A (BPA). *Reproductive Toxicology, 24*, 139–177.

Wang, Z., Zhang, X., Shen, P., Loggie, B. W., Chang, Y., & Deuel, T. F. (2006). A variant of estrogen receptor-{alpha}, hER-{alpha}36: Transduction of estrogen- and antiestrogen-dependent membrane-initiated mitogenic signaling. *Proceedings of the National Academy of Sciences of the United States of America, 103*, 9063–9068.

Waris, G., & Ahsan, H. (2006). Reactive oxygen species: Role in the development of cancer and various chronic conditions. *Journal of Carcinogenesis, 5*, 14.

Wehner, S., Buchholz, B. M., Schuchtrup, S., Rocke, A., Schaefer, N., Lysson, M., et al. (2010). Mechanical strain and TLR4 synergistically induce cell-specific inflammatory gene expression in intestinal smooth muscle cells and peritoneal macrophages. *American Journal of Physiology – Gastrointestinal and Liver Physiology, 299*, G1187–G1197.

Wu, R., Van der Hoek, K. H., Ryan, N. K., Norman, R. J., & Robker, R. L. (2004). Macrophage contributions to ovarian function. *Human Reproduction Update, 10*, 119–133.

Chapter 2
Cellular Plasticity, Cancer Stem Cells, and Cells-of-Origin

Outline

Introduction.. 21
Section 1: Defining the Terminology Necessary for Understanding This Field................... 22
Section 2: Confusion Among Stem Cell, Cancer Stem Cell, and Cell-of-Origin Concepts....... 26
Section 3: Evidence for the Theory of Cell-of-Origin as the Root of Tumors.................... 27
Section 4: Evidence for the Theory of Cellular Plasticity in Tumor Development............... 28
Section 5: Confounding Factors in the Plasticity Versus Cell-of-Origin Debate................ 28
References.. 30

Abstract It is now recognized that tumor cells share similarities to stem cells in that they do not have a strong sense of functional identity. In the case of the skin, a mature cell that becomes stem-like no longer has the identity of a skin cell, meaning it changes shape and behaves in ways that make it not fit in with its normal skin cell neighbors. Accompanying these changes in behavior are changes in which genes are transcribed or repressed. This is why the transcriptome of cancer cells or cancer stem cells often share many similarities with the transcriptome of embryonic or adult stem cells.

Introduction

What is the point of understanding whether a tumor arises from one cell or a number of different cells? These are the cells that allow the tumor to regrow after treatment has shrunk it. The regrown tumor is often meaner than the original tumor. These are also the cells the leave the tumor and colonize other organs in the form of metastasis, which is often what kills a patient. Thus, if medical research wants to develop more specific therapies that have fewer side effects and a lower risk of relapse, knowledge of cells-of-origin or cancer stem cells is essential. Furthermore, as nanotech research and in situ live-imaging technologies seek to treat cancer, they cannot be guided to target specific cells and spare normal cells if we do not know the properties of cells-of-origin or cancer stem cells.

© Springer International Publishing Switzerland 2016
D.H. Nguyen, *Systems Biology of Tumor Physiology*, SpringerBriefs
in Cancer Research, DOI 10.1007/978-3-319-25601-6_2

This chapter will discuss common concepts that are similar but not identical, which cause confusion within the field about how different studies relate to each other. For example, the cell-of-origin of a tumor may not be the same cell as the "cancer stem cell" (CSC) of the tumor. Furthermore, even if a tumor were derived from one transformed stem cell, there are processes within a tumor that would alter the identity of cells that arose from that initial stem cell, or that would produce hybrid cancer stem cells that may or may not resemble the initial stem cell.

Section 1: Defining the Terminology Necessary for Understanding This Field

To understand the concepts of cell-of-origin, cancer stem cell, dedifferentiation, and plasticity in cancer biology, it is important to clarify terms that are commonly used in the field and its various subfields.

Stem Cell

A stem cell is a cell that has two defining properties: (1) it can self-renew, meaning it can divide to produce two daughter cells, at least one of which remains a stem cell; and (2) it has the potential to mature (see the definition of "differentiation") into multiple different types of cells. When a stem cell divides to produce a stem cell and a progenitor cell, this is called *asymmetric division*. If a stem cell divided to produce two stem cells, then this is called *symmetric division*.

Progenitor Cell

A progenitor cell is similar to a stem cell, except that it has less potential with regard to what cell types it can mature to become. Depending on the type of tissue, progenitors can self-renew like stem cells do, producing more cells that maintain an identity of being a progenitor. Progenitors might also undergo what is called transit amplification, meaning they divide rapidly to form more copies of themselves. These types of progenitors are referred to as transit amplifying cells. The purpose of transit amplification is to produce many progenitors from a few starting progenitors, such that they turn into many mature cells. This effect spares stem cells from having to divide many times. For example, one stem cell can divide once to produce a stem cell and a progenitor cell. The resulting progenitor cell can then divide many times to produce many progenitor cells, so that the original stem cell doesn't need to divide unnecessarily. Depending on the organ, there can be early progenitors and late progenitors. The further down the line a progenitor becomes as it matures, the less potential it has to become different types of cells, meaning the more committed it is to a certain "fate," or identity.

Differentiation

Differentiation is the process in which a stem or progenitor cell matures into a cell with a specific function and identity. In cell biology and tissue biology, cellular function and identity go hand-in-hand. To speak generally, a stem cell that differentiates into one type of skin cell will take on the physical properties of skin and will not be able to do what a liver cell does for the body. Differentiation is how cells specialize in performing a certain duty, which requires them to turn on the genes and make the proteins necessary for that duty. The change in programing of what genes are turned on and which are turned off is an important reason why cells that differentiate into one type cannot differentiate into another type under normal circumstances. Differentiation is accompanied by a change in chromatin structure, epigenetic marks, transcriptomic profile, and cytoskeletal structure. A differentiated cell is a cell that has matured into a specific cell type that has a specific function.

Potency

This concept describes the potential that a stem or progenitor has for becoming different types of differentiated cells. The term *totipotent* means that a stem cell can become any differentiated cell type in the body. During mammalian embryonic development, a process called gastrulation occurs after which the three main germ layers are formed: endoderm, mesoderm, and ectoderm. Cells in these germ layer lineages are called *pluripotent*, meaning they can become multiple differentiated cell types that are derived from the same germ layer. For example, nerve cells, skins cells, and breast cells are all derived from the ectoderm. A *bipotent* progenitor cell is one that can only become one of two differentiated cell types within the same germ layer lineage. A *unipotent* progenitor is one that can divide to produce more of itself, but when its descendants differentiate, they can only become one cell type.

Lineage Commitment

This concept refers to the point of no return as a stem cell matures into an early progenitor and then to a late progenitor. The further along the progenitor goes along this continuum, the more it becomes committed to a certain fate, meaning the less potential it has to differentiate into different cell types.

Transdifferentiation

This describes the situation when a differentiated cell from one germ lineage (endoderm, mesoderm, or ectoderm) becomes a differentiated cell from another germ lineage. In mammals, this does not happen in a normal animal unless genetic engineering or tissue engineering is involved. Transdifferentiation is one of the main goals of adult stem cell research. For example, turning a person's skin cells (ectoderm origin), which are very abundant, into stem cells that can then be injected into that person's damaged heart (mesoderm origin) for the purposes of tissue regeneration.

Dedifferentiation

Dedifferentiation is the process in which a differentiated cell undergoes changes to lose its matured cell identity to become progenitor-like or stem-like. A dedifferentiated morphology is one of the hallmarks of cancerous cells. Dedifferentiation is accompanied by changes in cell-to-cell interactions, which allows the dedifferentiated cell to take on an odd shape or to divide "out of line" such that its daughter cells are no longer neatly packed like the other cells around them.

Plasticity

The concept of plasticity describes the potential of a differentiated cell to dedifferentiate back into a progenitor-like or stem-like state, and then to differentiate into a new differentiated state. The underlying concept is that a cell's identity, and thus its function, is not permanently fixed after a cell has differentiated. Within a tumor that contains many different regions of distinct morphology, or an early stage invasive tumor that is next to normal tissue, there are areas of transition showing groups of neighboring cells changing from one shape to another in ways that normal, differentiated cells do not do. Such drastic changes in cell morphology change the transcriptomic and proteomic identity of the cell. Thus, within the permissible environment of a tumor there is ample evidence of cellular plasticity.

Epithelial-to-Mesenchymal Transition (EMT)

Epithelial-to-mesenchymal transition (EMT) is a process in which an epithelial cell takes on the features of a mesenchymal cell, which allow it to break away from epithelial-type cell-to-cell interactions and to have enhanced migratory abilities. The function of epithelial cells is to form cube-like blocks that attached together to form a wall, which inherently limits their ability to move relative to mesenchymal cells that form layers of interspersed, flattened cells. The process of EMT has been well-documented to give an epithelial cell stem cell-like properties, though this is not necessarily true in every case

Niche

A niche is defined as the location in which a stem cell or progenitor cell likes to reside and perform its function of making progenitors or more of itself. The niche is a very important concept in stem cell biology and cancer biology, because those who study tissue development, organ regeneration, wound healing, or metastasis all study at least one facet of how the niche regulates stem cell function. Why are stem cells in tissue located where they are? What activates a dormant stem cell to produce progenitors that heal a wound? Do stem cell-like circulating tumor cells prefer certain locations when they seed into and metastasize in distant organs? Are these locations natural stem cell niches or does the roaming stem-like cell induce its own niche? Is there such a thing as an inducible niche? If so, how should we understand and predict sites of metastasis? These questions are highly relevant for understanding the fundamentals of organ development, wound healing, and tumor development. An organism's niche within its habitat is also an important concept in ecology. If defined in detail, the concept of a niche draws together many of the topics that are studied in ecology (i.e. types of species interactions, physical or biological factors that affect population density, ways of using available resources, waste management, mating habits, etc.). The niche is one of the concepts that unifies ecology. The niche is also a concept that unifies the topic of this chapter: plasticity, cancer stem cells, and cells-of-origin. A niche isn't just a basket of extracellular matrix proteins in which a stem cell sits. In a niche, the stem cell is surrounded by neighboring cells that communicate with it. These cells often play very important roles in regulating the activity of the stem cell. A stem cell in a niche can be affected by physical contact with its neighbors, paracrine signals released by its neighbors, signals from cellular projections that reach over from a distance, among other mechanisms of cell-to-cell communication. The reality, as with other aspects of biology, is likely multiple simultaneous signals.

Microenvironment

The microenvironment refers to physical attributes surrounding a cell, which includes the extracellular matrix molecules, temperature, pH, salinity, hormonal milieu, electro-magnetic signals, connective tissue cells (fibroblasts, blood vessels, lymph vessels), and immune cells that are present. The term microenvironment is often used to describe a general condition of, within, or surrounding a cell/tumor; often a condition that promotes a specific behavior of the tumor.

Clonality

This concept is important in discussions of cancer stem cells and cells-of-origin because it underlies part of the reason why certain therapies shrink a tumor, only to have the tumor return. A clone in this context is defined as a single cell that divides to give rise to a large population of identical cells. Polyclonal describes a tumor or population of cells that is derived from multiple distinct clones. Monoclonal describes a population of cells that is derived from one cell.

Cancer Stem Cell

A cancer stem cell (CSC) is a cell that divides to replenish a population of cancer cells. CSCs can exist in cell lines that are grown in two-dimensional and three-dimensional culture systems or in tumors that are grown in living organisms. CSCs are the cells that survive a therapeutic treatment that destroys the vast majority of the tumor. CSCs divide to produce cells that make up the returned tumor. CSCs have special properties that allow them to be distinguished from non-CSCs. CSCs are better able to repair damage that occurs inside of them, such as DNA and protein damage. They are also better able to quench reactive oxygen species inside of them, reducing the amount of damage that they incur when under stressful conditions. CSCs are often the cells that leave the primary tumor, survive in the blood stream, and metastasize distant organs. The terms cancer initiating cell (CIC) or tumor initiating cell (TIC) are often used to describe cells that are essentially CSCs, but are named differently because of how they were identified or tested as having stem cell-like properties. Section 2 of this chapter describes common confusions between CSCs and cells-of-origin.

Cell-of-Origin

A cell-of-origin is a normal stem or progenitor cell that becomes abnormal and gives rise to a cancer, liquid or solid. Due to its origin from a normal cell, the cell-of-origin is not necessarily the same thing as a cancer stem cell, though it may be. Cancer stem cells were derived from studying tumors, while cells-of-origin are identified by lineage tracing experiments that can track normal cells that become part of tumors. Section 2 of this chapter describes common confusions between CSCs and cells-of-origin.

Section 2: Confusion Among Stem Cell, Cancer Stem Cell, and Cell-of-Origin Concepts

The purpose of understanding cancers stem cells and cells-of-origin is to allow for the development of more specific therapies and therapies that result in less relapse. However, discussions of these two topics often results in confusion, since their definitions can overlap. Clarity on this matter is important because cancer prevention strives to keep normal cells from becoming cancerous, while cancer treatment seeks to eliminate cells that are already cancerous. The point of cancer research at the molecular and cellular level is so that these details allow us to produce better therapies that have fewer side effects. Clear definitions of function and identity are necessary for nanotechnology and targeted drug delivery methodologies to specifically target CSCs or cells-of-origin, while sparing normal cells.

The microenvironment of the tumor is drastically different than a normal tissue. Thus, even if a tumor arose from one cancerous cell among normal neighbors, by the time that one cell has divided enough times to form a tumor within that tissue, the microenvironment of that tumor may have turned non-CSCs into CSCs. Therefore, a tumor that started from a transformed normal stem cell, the cell-of-origin, may harbor CSCs that did not come from that stem cell. A tumor may harbor any combination of cells-of-origin and CSCs. This is because as a tumor grows, it enlarges and engulfs neighboring areas that also have normal stem cells. Those engulfed stem cells may then become additional cells-of-origin. Furthermore, the engulfed normal differentiated cells may become additional CSCs. Thus, a tumor may contain multiple cells-of-origin and multiple CSCs, each of which joined the tumor at different times. Given the complexity of the combinatorial possibilities of cells-of-origin and CSCs within a tumor, individual research papers need to assume simplifications in their conclusions about cells-of-origin and CSCs. However, for both those who are advancing this area of knowledge and those who are new to it, being aware of this complexity will help reveal new mechanistic insights, organize existing literature and future knowledge, and guide hypotheses.

Section 3: Evidence for the Theory of Cell-of-Origin as the Root of Tumors

Much has been written about the evidence for the cell-of-origin hypothesis. The reader is directed to the two excellent reviews that are briefly discussed here. The first is by White and Lowry (White and Lowry 2015), which discusses studies that provide evidence that specific populations within the hierarchy of stem cells and progenitor cells can be mutated to give rise to specific subtypes of cancers. The studies that were discussed yielded slightly different stem or progenitor cells as the cells-of-origin within hair follicles and the cells surrounding them. However, the clear uniformity from these studies is that causing mutations in a specific cell type within the stem cell hierarchy produces a specific type of tumor. While this supports the cell-of-origin theory, it is important to keep in mind that these studies are not yet able to trace the origin of the heterogeneous tumor—that is, one that has multiple distinct regions of morphology or behavior—to one or a few cells. The second review is by Jane Visvader (Visvader 2011), which outlines several conceptual models of how stem cells are mutated to become cells-of-origin. This review also discusses studies that support the cell-of-origin hypothesis. The best support comes from studies of hematopoietic cancers in which over expression of oncogenes in various compartments of the stem cell hierarchy resulted in cancer. The upper/earlier compartments, that is the stem cells, are more susceptible to oncogene-induced transformation that result in cancers, while the progenitor compartments are not. However, a higher dosage is able to produce cancer from the progenitor compartments, which highlights that given a permissive environment, such as a tumor, the right mutations can arise in cells that do not normally serve as cells-of-origin such that they become cells-of-origin. Further evidence comes from mouse models of intestinal cancer, prostate cancer, basal cell carcinomas, and pancreatic ductal cell carcinomas.

Section 4: Evidence for the Theory of Cellular Plasticity in Tumor Development

The cellular plasticity exhibited by partially or fully differentiated cells reverting back to stem cell is an alternative hypothesis about how tumors develop, especially those that harbor heterogeneous cell types. The liver is an organ that can regenerate itself after being wounded. Progenitor liver cells are believed to exhibit reversion back into stem cells even while in a non-cancerous—albeit wounded—organ state, in order to regenerate the region that is missing due to hepatectomy (Choi et al. 2014; Dorrell et al. 2011; Huch et al. 2013; Santoni-Rugiu et al. 2005). This type of plasticity is also observed in cancer. Klevebring and colleagues (2014) sequenced the exomes of CSCs and paired non-CSCs from breast cancer patients. To their surprise, the two populations shared the majority of the same somatic mutations, which argues against the idea that CSCs have a distinct mutational profile because they don't divide—and accrue replication errors—as much as the bulk of the tumor cells that arise from them. These data suggest that CSCs may be switching between a CSC and non-CSC state, since they share such similar somatic mutation profiles. Chaffer and colleagues (2011) made a discovery that also echoes the cellular plasticity theory. They found that differentiated human breast epithelial cells could spontaneously revert to a stem-like state. Furthermore, oncogenic transformation of these differentiated cells increased the rate of reversion to the stem-like state, producing CSC-like cells in vitro and in vivo. The plasticity of cancer cells is made extra evident in the study by Sharma and colleagues (2002), which showed that epithelial prostate cancer cells could dedifferentiate and then become vasculogenic cells that comprised blood vessels inside of tumors, something that epithelial cells are not supposed to do.

Section 5: Confounding Factors in the Plasticity Versus Cell-of-Origin Debate

It is difficult to clearly prove that a tumor arose from and is sustained by purely one cell-of-origin without the addition of CSCs that were derived from cells other than that cell-of-origin. If Sections 2 and 4 of this chapter did not make this point obvious, this section will further solidify it.

Cell Fusion

One major confounding factor in proving that a tumor is derived from one cell-of-origin is that spontaneous cell fusion can occur. Terada and colleagues found that mesenchymal stem cells from the bone marrow of mice could spontaneously fuse with mouse embryonic stem cells (Terada et al. 2002), which has deep implications for what the mesenchymal stem cells can do when they encounter CSCs or non-CSCs within the tumor microenvironment. Bone marrow-derived mesenchymal stem cells are common components of tumors. A study by Ying and colleagues further implicates the importance of cell fusion events that occur when stem-like cells interact. The authors isolated neural stem cells that were engineered to have permanent genetic markers (green-fluorescent protein and puromycin resistance), and then cultured them with mouse embryonic stem cells that were engineered to also have permanent genetic markers (hygromycin resistance). They discovered that the neural stem cells could spontaneously fuse with the embryonic stem cells to produce hybrid cells that contained genetic markers from both cell types (Ying et al. 2002). The publication of the Terada et al. and Ying et al. papers required a reinterpretation of a number of transdifferentiation studies in which cell fusion may actually have been the mechanism, instead of transdifferentiation, by which the cells' phenotypic identities were altered (reviewed in (Wurmser and Gage 2002). In 2013, Lazova and colleagues reported the first case of cell fusion in human cancer. By measuring regions of DNA that were unique to the recipient or donor from whom bone marrow was taken, the authors were able to show that the melanoma that metastasized to the brain of the recipient was a hybrid of both the recipient and the donor (Lazova et al. 2013).

Induced Mesenchymal States

Epithelial-to-mesenchymal transition (EMT) is a process in which an epithelial cell takes on the features of a mesenchymal cell, which allow it to break away from epithelial-type cell-to-cell interactions and to have enhanced migratory abilities. The function of epithelial cells is to form cube-like blocks that attach together to form a wall, which inherently limits their ability to move relative to mesenchymal cells that form layers of interspersed, flattened cells. The process of EMT has been well-documented to give epithelial cells stem cell-like properties, though this is not necessarily true in every case (reviewed in (Brabletz 2012)). Much has been written about growth factors and over-expressed genes that induce an EMT and CSC phenotype.

This section will discuss a mechanism of cell-to-cell communication that allows for the transfer of nucleic acid polymers and it's implication on bulk tumor gene expression profiling data. Exosomes are microscopic membrane vesicles that bud off from a donor cell and that can fuse with another cell, dumping their contents into the recipient cell. The discovery that exosomes can deliver nucleic acid polymers, such as microRNA (miRNA), which can alter the transcript level of their target genes in the recipient cell (Melo et al. 2014; Singh et al. 2014; Yang et al. 2011) has important implications for understanding mechanisms that induce cellular plasticity. These mechanisms of nucleic acid transfer make sense of why tissues that seem morphologically normal to a pathologist can have a stem cell-like or progenitor-like transcriptomic signature that is detected by mRNA extraction from bulk tissue. Spatially guided, laser-microdissection transcriptomic techniques that allow the investigator to identify the transcriptome of distinct regions or cells within a sample may clarify epicenters of exosome-mediated nucleic acid transfer (reviewed in (Crosetto et al. 2015)).

References

Brabletz, T. (2012). EMT and MET in metastasis: Where are the cancer stem cells? *Cancer Cell, 22*, 699–701.

Chaffer, C. L., Brueckmann, I., Scheel, C., Kaestli, A. J., Wiggins, P. A., Rodrigues, L. O., et al. (2011). Normal and neoplastic nonstem cells can spontaneously convert to a stem-like state. *Proceedings of the National Academy of Sciences of the United States of America, 108*, 7950–7955.

Choi, T. Y., Ninov, N., Stainier, D. Y., & Shin, D. (2014). Extensive conversion of hepatic biliary epithelial cells to hepatocytes after near total loss of hepatocytes in zebrafish. *Gastroenterology, 146*, 776–788.

Crosetto, N., Bienko, M., & van Oudenaarden, A. (2015). Spatially resolved transcriptomics and beyond. *Nature Reviews Genetics, 16*, 57–66.

Dorrell, C., Erker, L., Schug, J., Kopp, J. L., Canaday, P. S., Fox, A. J., et al. (2011). Prospective isolation of a bipotential clonogenic liver progenitor cell in adult mice. *Genes and Development, 25*, 1193–1203.

Huch, M., Dorrell, C., Boj, S. F., van Es, J. H., Li, V. S., van de Wetering, M., et al. (2013). In vitro expansion of single Lgr5+ liver stem cells induced by Wnt-driven regeneration. *Nature, 494*, 247–250.

Klevebring, D., Rosin, G., Ma, R., Lindberg, J., Czene, K., Kere, J., et al. (2014). Sequencing of breast cancer stem cell populations indicates a dynamic conversion between differentiation states in vivo. *Breast Cancer Research, 16*, R72.

Lazova, R., Laberge, G. S., Duvall, E., Spoelstra, N., Klump, V., Sznol, M., et al. (2013). A melanoma brain metastasis with a donor-patient hybrid genome following bone marrow transplantation: First evidence for fusion in human cancer. *PLoS One, 8*, e66731.

Melo, S. A., Sugimoto, H., O'Connell, J. T., Kato, N., Villanueva, A., Vidal, A., et al. (2014). Cancer exosomes perform cell-independent microRNA biogenesis and promote tumorigenesis. *Cancer Cell, 26*, 707–721.

Santoni-Rugiu, E., Jelnes, P., Thorgeirsson, S. S., & Bisgaard, H. C. (2005). Progenitor cells in liver regeneration: Molecular responses controlling their activation and expansion. *APMIS (Acta Pathologica, Microbiologica et Immunologica Scandinavica), 113*, 876–902.

Sharma, N., Seftor, R. E., Seftor, E. A., Gruman, L. M., Heidger, P. M., Jr., Cohen, M. B., et al. (2002). Prostatic tumor cell plasticity involves cooperative interactions of distinct phenotypic subpopulations: Role in vasculogenic mimicry. *The Prostate, 50*, 189–201.

Singh, R., Pochampally, R., Watabe, K., Lu, Z., & Mo, Y. Y. (2014). Exosome-mediated transfer of miR-10b promotes cell invasion in breast cancer. *Molecular Cancer, 13*, 256.

Terada, N., Hamazaki, T., Oka, M., Hoki, M., Mastalerz, D. M., Nakano, Y., et al. (2002). Bone marrow cells adopt the phenotype of other cells by spontaneous cell fusion. *Nature, 416*, 542–545.

Visvader, J. E. (2011). Cells of origin in cancer. *Nature, 469*, 314–322.

White, A. C., & Lowry, W. E. (2015). Refining the role for adult stem cells as cancer cells of origin. *Trends in Cell Biology, 25*, 11–20.

Wurmser, A. E., & Gage, F. H. (2002). Stem cells: Cell fusion causes confusion. *Nature, 416*, 485–487.

Yang, M., Chen, J., Su, F., Yu, B., Su, F., Lin, L., et al. (2011). Microvesicles secreted by macrophages shuttle invasion-potentiating microRNAs into breast cancer cells. *Molecular Cancer, 10*, 117.

Ying, Q. L., Nichols, J., Evans, E. P., & Smith, A. G. (2002). Changing potency by spontaneous fusion. *Nature, 416*, 545–548.

Chapter 3
Using Mouse Models and Making Sense of Them

Contents

Introduction.. 33
Section 1: Selecting the Right Model.. 35
Section 2: Correctly Using the Right Model... 40
Section 3: Making Sense of the Data... 44
References.. 51

Abstract The purpose of this chapter is to provide some insights into the use and misuse of mouse models of cancer. The goal is that this chapter will help investigators better plan their mouse experiments. It also will be helpful in making sense of data that is conflicting or incongruous across independent experiments or research groups. This chapter is structured into three sections to give coherence among the individual topics that are discussed. The sections also represent the three general phases of a study involving rodent models and the challenges that are commonly encountered in each. The guidelines are certainly not exhaustive, since there are many factors that were not discussed due to space limitations.

Introduction

Section 1: Selecting the Right Model
Section 2: Correctly Using the Right Model
Section 3: Making Sense of the Data

In vivo experiments have an unpredictable nature to them, which underscores the complexity of mouse physiology that is yet to be completely understood. While there are no conventions that every research group follows when performing mouse experiments, appreciating the complexities highlighted in this chapter will make it easier to compare independent data sets in a meaningful way. There has been an effort to systematically define conventions for reporting the experimental details and results of rodent models: ARRIVE (www.nc3rs.org.uk/ARRIVE) (Kilkenny et al. 2010). Other institutions are attempting to provide frameworks for systematically reviewing rodent data: CAMARADES (www.camarades.info) (Sena

© Springer International Publishing Switzerland 2016   33
D.H. Nguyen, *Systems Biology of Tumor Physiology*, SpringerBriefs
in Cancer Research, DOI 10.1007/978-3-319-25601-6_3

et al. 2014), SYRCLE (www.syrcle.nl) (Hooijmans and Ritskes-Hoitinga 2013), and SABRE (http://www.sabre.org.uk/) (Muhlhausler et al. 2013). These prescribed conventions are an important move in the right direction, since their goal is to improve transparency, inter-study comparability, and reproducibility. The conventions should be increasingly enforced by funding agencies and journals for the following reasons. (1) Meta-analysis of pre-clinical rodent models can identify adverse effects of novel treatments before those treatments go into clinical trials on humans (reviewed in (Pound et al. 2004)). (2) More reliable studies prevent biomedical research involving animal models from losing credibility as a worthwhile scientific pursuit. (3) More reliable studies keep tax-payer-funded researchers accountable to the public, which will eventually get fed-up with bad subfields of science. (4) More reliable studies reduce the waste of tax-payer funds and curtail the risk of further reduced funding for certain subfields of research.

Many of the topics discussed in this chapter are factors that increase the "noise" within data from mouse experiments. While there is no way to completely control every single factor, a few simple precautions can dramatically increase the yield of "clean" data. Studies involving mouse models often become tissue banks of information that can be probed to find new insights or directions for research. Taking the guidelines outlined in this chapter into consideration while planning and doing mouse experiments may help the investigators get more mileage out of their data.

There is an increasing distrust of research involving animal models, especially in areas that seek to find new treatments that will immediately translate into treatments for human diseases (Hayden 2014; Lutz 2011; SABRE-Research-UK 2014). There are several reasons behind this frustration. (1) Studies using inbred mouse strains do not represent the genetic diversity that exists in human populations, so the effectiveness of mice as a model of a human disease may need to be tested across multiple different strains, depending on the question of interest. (2) Many animal studies are often not done very well, as is commonly stated in the publications that systematically review animal model data prior to, or in light of, human clinical trials that fail (Hooijmans and Ritskes-Hoitinga 2013; Kilkenny et al. 2010; Muhlhausler et al. 2013; Sena et al. 2014). (3) A mouse is physiologically not the same as a human—it's just a model, not a replica. (4) The physiology of a mouse is very complicated and we have yet to understand enough that we can control the experiment to the degree that we want. In light of these challenges, this chapter attempts to help make things clearer along the lines of the book's subtitle: "Rethinking the Past, Defining the Future." It does so by suggesting guidelines that may shed light on previous data while guiding the structure of future experiments such that more useful information can be extracted within and across independent studies. The guidelines in this chapter also follow the theme of the book's main title: "Systems Biology of Tumor Physiology." The guidelines treat the tumor as if it were an endocrine organ that communicates with the brain and other endocrine organs. The guidelines suggest that more information can be gained from a mouse model of cancer if we pay attention to the global physiology of the mouse while considering the cancer as an organ that affects, and is affected by, that physiology. The guidelines in this chapter come from personal experience, having wrestled with the literature, peer-reviewing literature, conversations, and commiserations with scientists from across the globe.

Section 1: Selecting the Right Model

The Age of the Mice

Mice undergo developmental stages as they mature, just as humans do. Mice begin puberty at around 3–4 weeks after birth and reach young-adulthood at 8–9 weeks of age. In females, the mammary gland has fully developed, filling the mammary fat pad, by the 8–9 week time point (Sternlicht 2006). Organs undergo many dramatic changes during puberty and continue to do so through the lifespan of the mouse. Organs physiologically and genetically change when transitioning from puberty, to adulthood, to pregnancy, and to old age. Mice typically have a life span of 2–3 years. A 1-year old mouse can be considered as a "middle-aged" mouse. Two-year old mice are considered "old." Investigators who are interested in studying age-related effects should be cognizant of the definitions of age ranges within the literature, as this can be a source of confusion across independent studies.

Synchronizing the age of the mice at the start of the experiment and keeping track the date and time of procedures during the experiment will become helpful in making sense of the results. Synchronization here means selecting mice based on the timing since birth and weaning. This information may explain differences between experiments past and present, within one research group or across multiple research groups.

An important factor to also keep in mind is that mice of the same age also vary in weight, since a litter of pups may have a few runts that are smaller than their siblings. Thus, when purchasing mice of a certain age from a vendor, it will be helpful to request that the mice be above a certain weight, not just at a certain age. This simple step will help reduce variation in the data. The optimal weight must be determined by the investigator. This factor about age and weight also applies to research groups that breed their own mice.

Appreciate the Estrus Cycle to Reduce Variation in Data

The estrus cycle in mice occurs in cycles of 4–5 days, depending on the strain. It is composed of four phases (Byers et al. 2012): proestrus, estrus, metestrus, and diestrus. The first two phases, proestrus and estrus, is when the female mouse is most receptive to mating and pregnancy. The second two phases, metestrus and diestrus, is when the uterine lining degrades. The hormonal milieu of these four phases is dramatically different. Since the endocrine hormones from the brain and the ovaries surge throughout the blood stream, the different levels of hormones during each stage may confound data, adding extra "noise" by causing a wide variation. Estrus stage should be considered when the scatter of data points within a treatment group is wide, exhibiting a bimodal distribution of two clusters separated by a few data points in between them. The average value of the data points of the bimodal distribution may be the same as that of the unimodal distribution in the control group, but annotating the data points according to estrus stage may reveal that there are two distinct biological responses in the bimodal distribution.

Some investigators synchronize their female mice several days before a treatment begins by injecting a cocktail of estrogen and progesterone under the skin between the scapulae. This is to avoid the confounded effects of having multiple different estrus stages present at the time that a treatment is administered.

The Estrus Cycle Affects the Immune System

The estrus cycle affects many organs and systems in the mouse because many different tissues have protein receptors for the endocrine hormones that regulate and facilitate estrus cycling. This includes cells in the innate or adaptive immune systems (Beagley and Gockel 2003; Chua et al. 2010; Petrovska et al. 1996). The immune system has intricate functions in each organ system. This is part of the reason why estrus staging may be the culprit behind the wide scatter of data within a treatment group.

Identifying Estrus Stage

It is good practice to record the estrus stage at the time that a treatment is administered to the mice. It is also good practice to record the estrus stage at the time that tissue is harvested from the mice. The study by Byers and colleagues (2012) describes an easy method for determining estrus stage by visually inspecting the vaginal orifice. The authors also provided pictures of the orifice at each stage for three strains of mice: BALB/cByJ (white coat), CByB6F1/J (agouti coat), and C57BL/6J (black coat). The authors explain that visual inspection is the easiest and quickest method to stage for estrus, but is only reliable in identifying the first two stages of the cycle, proestrus and estrus. However, having only this minimal information as part of the data records may turn out to be very useful. The most reliable method for identifying each of the four stages of estrus is the cytology method in which a moist cotton swab is used to gently scrape cells from the vaginal wall. The study by Byers et al. also provides reference images for cell morphologies at each stage and provides citations for more detailed explanations.

Daily Estrus Staging Can Cause Pseudopregnancy

Daily staging of estrus via physical entry into the vagina may cause a mouse to become pseudopregnant. When a female mouse mates with a sterile male mouse, the female mouse becomes pseudopregnant. Her internal hormones change as if she was pregnant, her mammary glands proliferate and produce milk, and she builds a nest. However, she is not pregnant and will not deliver pups. Daily staging of estrus can induce pseudopregnancy, which changes the hormonal milieu of the data that is subsequently collected. This should be kept in mind when interpreting data from experiments that performed daily invasive staging. A less invasive, but more costly, means of daily estrus staging is the collection and analysis of urine. In mice, the fourth pair of mammary glands, known as the inguinal glands, is often manipulated during the experiment due to ease of surgical access. Preserving the fourth pair for manipulation, the third pair of mammary glands under the upper limbs can be collected, fixed and stained by the "whole mounting" process (Plante et al. 2011), and examined to determine if a state of pseudopregnancy is present. Pseudopregnancy, like pregnancy, results in engorged ducts and lobules in the mammary gland, which fills much of the fat pad (Brisken 2013; Oakes et al. 2006).

Background Genetics

Genetically engineered mice may exhibit odd mating phenotypes. They may also develop cancers in tissues other than at the site of transplantation or interest. Sometimes this is due to the "leakiness" of the artificial promoter that controls the expression of the transgene in a genetically modified mouse. It is worth noting that certain strains of mice are genetically prone to reproductive problems, which may be amplified by the fact that they have been genetically modified by the insertion or removal of genetic material.

The results of a mouse experiment may be more dramatic if the experiment is performed on an inbred mouse strain that has certain genetic susceptibilities. Mouse models are just that, research models. As someone once insightfully stated, "No model is right, but some are useful." Mouse models are used to show proof of principle about a biological mechanism of action. Whether the mouse experiment showed no effect or a big effect, it's still just a mouse model that is different from humans. Aside from situations in which the researcher wants to test the effect of genetic background, researchers should make the most out of their experiments and pick a mouse strain that will give them the clearest data possible. The most rigorously supported claims—usually from multiple independent publications by independent laboratories—that are based on mouse experiments are those in which the evidence has been independently repeated in multiple strains of mice. This is not to say that a study that only tested one mouse strain is poorly designed, since testing multiple mouse strains in one publication is often financially unfeasible for one laboratory. Furthermore, without a clear phenotype from a susceptible inbred strain, it would be difficult to identify a cellular or molecular mechanism that is responsible for that phenotype and that might translate to a treatment for the human disease that is being modeled.

There is abundant evidence that cancer phenotypes are dramatically affected by the strain of the mouse. A first example comes from the study of the polyoma virus. This virus infects mice and causes them to develop mammary cancer. However, different inbred strains have different susceptibilities to polyoma virus-induced mammary cancer: C3H/DiBa mice are highly susceptible; DBA/2 and BALB/c mice have low susceptibility, while C57bl/6 mice are resistant. It was found that the H-2 haplotype, which is the major histocompatibility complex (MHC) gene, of the strain correlated with the degree of susceptibility, suggesting that immune resistance as a major mechanism behind these phenotypes (Freund et al. 1992). A second example of the influence of background genetics on cancer phenotype is a study of the Polyomavirus Middle T Antigen (PyMT) mouse model. PyMT mice harbor the middle T antigen of the polyoma virus, which causes mammary cancer. The investigators compared the cancer phenotype of two strains of PyMT mice, FVB/NJ and C57bl/6J. The PyMT model is a very aggressive form of mammary cancer in the FVB strain, but the same PyMT oncogene became less aggressive in the C57bl/6J strain (Davie et al. 2007).

A third example of the importance of genetic background on cancer susceptibility was revealed by studies that removed the *Trp53* gene from mice. The protein product of this gene is p53, a transcription factor that is involved in many cellular functions, ranging from DNA damage response (Reinhardt and Schumacher 2012), to metabolism (Vousden and Ryan 2009), and to stem cell regulation (Spike and Wahl 2011). The *Trp53* gene was first removed in a mouse that was a cross of the C57bl/6 and 129/Sv backgrounds (C57bl/6 × 129/Sv) (Donehower et al. 1992). These mice began to develop tumors at around 4.5 months of life. When these null mice were crossed into a pure 129/Sv background, the descendants not only developed tumors sooner, but had testicular tumors 50 % of the time. Since the 129/Sv background had a natural susceptibility to testicular cancer, it appeared that removing the *Trp53* gene enhanced the tendency of these mice to form this cancer (Donehower 1996). In contrast, the C57bl/6 strain had a natural susceptibility to lymphomas (Stutman 1974), which was enhanced when *Trp53* was removed. A similar effect was observed when the *Trp53* null allele was crossed into the BALB/c background. The BALB/c strain had a natural susceptibility for mammary cancer, and removing the *Trp53* gene enhanced this phenotype (Jerry et al. 2000; Kuperwasser et al. 2000).

Single Nucleotide Polymorphism (SNP) arrays can be useful for reconciling or justifying, phenotypic differences across experiments, but more cost effective methods, such as histology, may also be useful. The topic in this chapter entitled "Collect the Endocrine Organs for Retrospective Analysis" discusses the utility of harvesting endocrine organs at the time of euthanasia for the purposes of making sense of perplexing data. This topic is also relevant for reconciling effects that are due to genetic background. Collecting organs that are known to express a disease phenotype that is associated with a certain mouse strain may turn out to be quite useful. Histological and/or "omic" analyses of these organs may explain differences in phenotypes observed in certain mouse strains across independent experiments. It is helpful to keep in mind that different venders that provide the same strain may have derived that strain from different sub-strains. If collaborating laboratories followed a convention to harvest and store the same endocrine organs from 5 to 10 mice per treatment group in an experiment, these tissues may become very helpful in reconciling conflicting or incongruous data.

Also see the topic entitled "Developmental Deformities Give Insight into Future Phenotypes" in Sect. 1 of this chapter.

Developmental Deformities Give Insight into Future Phenotypes

Certain ailments and physical deformities are common to laboratory mice. The chapter by Burkholder and colleagues entitled "Health Evaluation of Experimental Laboratory Mice" provides a good synopsis of common health problems in laboratory mice (Burkholder et al. 2012). Understanding these ailments is not just important for humane mouse husbandry, but helps identify clues about incongruous phenotypes between independent research groups. Understanding the genetic susceptibility of parental mouse strains, such as the high incidence of scrotal hernias in the FVB strain of mice (Lewis et al. 2012), may help investigators understand phenotypes such as low fertility rates in their particular genetically engineered mice. Keeping the knowledge of genetic susceptibilities in mind, or knowing to look for them in the literature, will save time and effort when overcoming sudden problems such as low birthrates and infertility. The background susceptibility interacts with the genetic modification to yield abnormal reproductive phenotypes that can delay the preparation process before the experiment even begins.

The nature of a common deformity can give insight into why a certain mouse model yields a certain experimental result while other similar models give similar, but slightly different results. These differences may be things like a different spectrum of tumor types or tumor subtypes between similar mouse models, or more of a certain phenotype (i.e. more squamous carcinomas in one strain compared to a related, but mixed strain; or more bone metastasis in one strain compared to a related strain). It is useful to keep in mind that multiple different adult tissues arose from the same embryonic germ layers (MacCord 2013; Tam and Beddington 1992): endoderm, mesoderm, and ectoderm. In mice and humans (and other mammals), both skin cells and mammary cells originated from the ectoderm layer. Thus, skins cancers and mammary cancers, which can revert to an embryonic-like state, may share more similarities to each other than to osteosarcomas. This is partly because osteosarcomas arise from bone, which originated from a different germ layer (the mesoderm) than the skin and mammary gland.

Understanding the relationships between germ layers and adult tissue programing will guide what questions to ask of the literature and of the data when seeking to make sense of similar themes between distinct diseases. As a hypothetical example, noticing that about 10 % of the mice have skin lesions or a skin deformity may shed light on differences in the mammary tumor data of this study compared to that of another study that bred its own mice. Long-term breeding of mice by individual research groups may result in genetic drift, producing sub-strains of the original strain. This inadvertently enriches for common deformities that are very rare in the original inbred strain.

The Composition of Mouse Chow

Soy-based mouse chow contains phytoestrogens that have endocrine effects. Investigators who study tissues that are known to be responsive to estrogens and progestins should be aware of the composition of the chow that is fed to their mice. Degen and colleagues measured the amount of genistein and daidzein, the two main soy isoflavones, in various rodent chows (Degen et al. 2002). Soy-based chows can have dramatic effects on phenotypes of both male and female mice (Aukema et al. 1999; Quiner et al. 2011; Stauffer et al. 2006). The composition of chow may be a commonly overlooked confounding factor in rodent studies. Thigpen and colleagues discuss factors to consider when choosing the appropriate rodent diet for studies of endocrine disrupting agents (Thigpen et al. 2004).

Section 2: Correctly Using the Right Model

Mock Procedures for Control Groups Yield Cleaner and More Consistent Data

Experiments involving mouse models need negative controls just like other types of experiments. It is important to keep in mind that taking a mouse out of its cage, physically restraining it, and subjecting it to a foreign chamber, surgery, or injections is a very stressful process for the mouse. Psychological and physical stress alters the physiology of the mouse. Thus, it is important that the negative control groups be subjected to the same procedures as the experimental group of mice. Performing a mock procedure will subject the control group to the same stressful situations that the experimental group must endure. For example, experiments that perform surgery on the experimental group should also perform the same anesthetization procedure and similar surgery on the control group. Experiments that involve injections of compounds into the experimental group should also inject the liquid vehicle, but without the compound, into the control mice (the "vehicle" is the liquid in which a compound is dissolved). Experiments that involve irradiating mice in a metallic X-ray machine that has a heavy slamming door and that makes loud humming noises should put the control mice through the same procedure but without turning on the radiation.

Different Types of Stressors

Various forms of stress will alter the physiology and gene expression patterns of mouse tissues. These stressors include audiogenic stress (noise, predator sounds within an animal facility that houses multiple species), physical stress (restraint, physical harm), toxicological stress (toxins, chemicals, radiation), and social stress (being caged alone, being caged with an aggressive mouse, not being able to burrow and hide). Studies about the effect of chronic mild stress in mice revealed many layers of complexity regarding how mice respond. The effect of this chronic mild stress depends on the type of stressor, the pattern in which the stress is applied (predictable or unpredictable), and the strain of mice that was studied (Ducottet et al. 2004; Ducottet and Belzung 2004; Griffiths et al. 1992; Ibarguen-Vargas et al. 2008; Mineur et al. 2006; Pothion et al. 2004). Furthermore, picking up and restraining a mouse for the purpose of an intraperitoneal injection is itself a stressful experience for the mouse, which is why mice in control groups should also undergo the process of being picked up, restrained, and injected with the vehicle solution. The routine handling of mice—picking up the mouse, holding it in your hands—can increase their stress response (Meijer et al. 2007). It is worth noting that the gender of the experiment has a significant effect on a mouse's stress response. Male humans and other mammals, not the females, produce a volatile chemical that can be smelled by the mice. This male olfactory signal triggers an increase in the stress hormone corticosterone, which dampens the amount of pain that the mice feel after receiving in injection in an ankle. Thus, it is worth keeping track of and reporting the gender of the experimenters and when in the experiment a change in experimenters occurred (Sorge et al. 2014).

Environmental Enrichment and Cancer

A study by Cao and colleagues found that providing mice with environmental enrichment (EE) reduces tumor size, tumor growth rate, and tumor frequency across multiple types of cancers (Cao et al. 2010). EE is a spacious housing condition that includes toys for mice to play in/with and material with which they can build nests. However, the results of the Cao et al. study could not be replicated by an independent research group (Westwood et al. 2013), which further underscores the complexity of mouse experiments and the need for guidelines such as those outlined in this chapter.

Two other studies about EE on cancer remain, but like the Westwood study (Westwood et al. 2013) neither reported that EE resulted in smaller tumors as was observed in the Cao study. In the first remaining study, Benaroya-Milshtein and colleagues showed that EE causes mice to have slower tumor growth after injection with an idiotype-vaccine prior to injection of tumor cells. The tumor-inoculated mice housed in EE lived longer and 44 % were tumor-free, compared to tumor-inoculated mice housed under standard conditions (Benaroya-Milshtein et al. 2007). In the second remaining study, Nachat-Kappes and colleagues showed that EE resulted in a slower tumor growth rate, but only up to 10 days after injection of the tumor cells. Tumors from mice with EE had reduced levels of COX-2 and Ki67, but higher levels of caspase-3 (Nachat-Kappes et al. 2012). In summary, these studies show that EE does have biological effects on mouse physiology and tumor pheno-type, though none of the studies show a consistently dramatic effect of EE on tumor size. Further studies are warranted, especially with a concerted effort to make EE parameters uniform across the studies.

Also see the topic entitled "Collect the Endocrine Organs for Retrospective Analysis" within Sect. 3 of this chapter.

Time of Day for Treatments and Exposures

Repeated treatments, exposures, or surgeries should be done at the same time of day each time. This will minimize the phenotypic variations that result from differential physiological states due to circadian rhythms. Nearly all species have adapted to earth's 24-h rotation, known as the circadian rhythm. Alterations in this rhythm due to sleep disorders, exposure to artificial light, exposure to electromagnetic radiation, and shift work have been linked to many metabolic and physiologic diseases (reviewed in (Takahashi et al. 2008)). The genes PER1 and PER2 are activated in the brain after exposure to pulses of light. A handful of other genes (CLOCK, CRY, BMAL, NPAS, and DEC) and their isoforms interact with the PER genes in a feed-back loop that is regulated by light and the autonomic nervous system (reviewed in (Sahar and Sassone-Corsi 2009; Savvidis and Koutsilieris 2012)).

Mouse physiology during the mornings is different than that during the afternoon or evening. This means that how a mouse's physiology responds to a drug, a wound, or a stressful situation differs depending on what time of the day it is. I was told of a case in which injecting a normally innocuous chemical into mice killed them, but that this lethal effect could never be repeated by animal welfare inspectors. It turned out that the chemical caused lethality when injected before noon time, and that the inspectors habitually did their inspection in the afternoons when the mice's physi-ological state could tolerate the chemical. Indeed, the "time of day" at which proce-dures were done is one of the details on the "ARRIVE guidelines checklist" for reporting animal studies (http://www.nc3rs.org.uk/page.asp?id=1357).

The Confounding Effects of Anesthesia

The confounding effects of anesthetics are especially important for prolonged tumor studies during which the mice are repeatedly anesthetized for surgical or biopsy procedures. A wide array of chemicals can be used to anesthetize mice. For tumor biology experiments, it is helpful to know which anesthetics have been shown to have anti-tumor properties. For example, the analgesic Meloxicam has been reported to inhibit the growth of several different cancer cell lines in culture and of ovarian cancer subcutaneous xenografts (Ayakawa et al. 2009; Goldman et al. 1998; Xin et al. 2007). Thus, searching for literature about an anesthetic of interest may allow the researcher to avoid confounding factors in mouse experiments. Anesthetics that come in the form of volatile inhalants have been shown to alter the gene expression of human cancer cell lines. The human breast cancer cell line MCF7 exhibited a different kinetic of gene expression change over a 1-h time course than did the human neuroblastoma cell line SH-SY5Y. Different types of volatile inhalants induced different kinetics of gene expression change (Huitink et al. 2010).

All analgesics can have confounding effects. The nature of science rarely allows a researcher to completely nullify all confounding factors, but safeguards can be implemented for the attainment of cleaner data. In fact, understanding how confounding factors affect data is one way by which new discoveries are made. Different classes of analgesics have different mechanisms of action. Here are some questions that will help you decide if a certain anesthetic should be avoided. (1) Are the chemical's mechanisms of action known to be highly active in the general phenotype of your research model? (2) Is the protein target of the anesthetic highly expressed in your research model? (3) If you are comparing two or more subtypes of a cancer, does one of them express the anesthetic's inhibitory target more than the other? The answers to these questions may not be available before the experiment is done, but can be obtained after tissues have been harvested. The answers may come in handy when interpreting the data that is inconsistent with those from other studies or suddenly inconsistent when compared to previous studies.

Examples of Rodent Anesthetics

Inhalant anesthetics include isofluorane, halothane, sevoflurane, methoxyflurane, nitrous oxide, and ether. Anesthetics combined with Ketamine: Xylazine, Acepromazine, Medetomidine, Midazolam. Other anesthetics are Buprenorphine is an opiod analgesic used for pre- and post-operative purposes. Non-steroidal anti-inflammatory (NSAID) analgesias include Carprofen, Meloxicam, Ketaprofen, Ketorolac, and Flunixin meglumine. Local analgesics used on the area of surgical incision include Lidocaine hydrochloride and Bupivacaine. For more information on the use of these compounds see IACUC (2011).

Completely Thaw Frozen Plasma and Re-Suspend for Cleaner Data

Many studies collect blood plasma for analysis of proteins, metabolites, etc. Plasma is often frozen at minus 80 °C for later analysis. Plasma is a complex mixture of proteins, carbohydrates, and metabolites. The components in plasma have different densities and will freeze and thaw at different rates. Therefore, it will be advantageous to make sure frozen plasma is completely thawed on ice and re-suspended with a micropipette before aliquots are taken for analysis. This simple step will yield cleaner, more repeatable data.

Section 3: Making Sense of the Data

Collect the Endocrine and Other Organs for Retrospective Analysis

Endocrine systems are integrated and communicate with each other via positive and negative feedback loops. Harvesting tissues in a small population of mice (5–10 mice per group) from each experimental condition will provide useful information about the global physiological status of the mice. The information from organs outside of the tumor will provide a contextualized picture of why a tumor is behaving the way it does in one treatment but not another. A tumor can be considered an organ not just because it shares morphological similarities with the tissue from which it arose, but also because, like organs do, it communicates with other organs via endocrine pathways.

The following items are examples of ways in which a cancer's effect and behavior can be better understood by quantifying the characteristics of organs that it affects. *These surrogate metrics are not substitutes for direct measurements on a tumor, but together with the direct measurements on a tumor they provide a holistic understanding of the complex, integrated mechanisms that yield a tumor's phenotype. These surrogate metrics are also great for (1) generating mechanistic hypotheses, and (2) comparing incongruent mouse data sets that should otherwise be similar.* The following examples were selected because they can be measured with a ruler or weight scale, but which organs to pick and how to measure them (i.e. non-invasive imaging) depends on the details of the experiment and the available technologies. Organs can be fixed in formalin or other appropriate fixatives for long-term storage and later analysis. Or, the organs can be measured and then immediately processed for extracting live cells. Preserving an organ by formalin-fixation is advantageous in that it allows paraffin-embedding, slicing, and histological examination. Histology can provide much more information than size and weight alone.

The Thymus

The thymus is the organ where T lymphocytes mature. It grows rapidly after birth, but begins to shrink at the start of puberty. However, when a mouse is stressed, the thymus shrinks faster than usual (Dominguez-Gerpe and Rey-Mendez 2003; Pearse 2006). Thus, thymus weight and size may serve as a surrogate measure of inadvertent stressors that affected one treatment group but not another. Normal, age-related shrinkage of the thymus is considered involution, while stress-induced shrinkage of the thymus in young-adult mice is considered shrinkage. It should be noted that in old mice, involution and atrophy may appear histologically similar (Pearse 2006).

The Spleen

An enlarged spleen is indicative of an activated immune system (Bronte and Pittet 2013). A difference of weight and size of the spleen between two experimental conditions suggest that one condition is activating the immune system. Without further dissection of cell surface markers and functional in vitro assays on primary splenic lymphocytes, it is difficult to pinpoint a more detailed mechanism. However, an enlarged spleen is indicative of an immune defense against factors that the mouse considers as harmful. Measuring the spleen at the time of euthanasia is a simple surrogate measure of differential immune activity across experimental groups of mice.

The Liver

The liver is responsible for a number of homeostatic functions, including detoxification and response to acute infection. Part of the liver's response is to increase in size, which it does in response to the aforementioned insults. However, it also increases in size in response to normal physiological hormones such as those governing pregnancy and lactation; and it grows in response to an increased dietary intake of fat, carbohydrates, and protein (Maronpot et al. 2010). Thus, the size and weight of the liver is a surrogate measure of many potentially overlooked physiological mechanisms underlying a cancer phenotype. It may be useful in reconciling incongruent data between mouse experiments that should have exhibited similar results.

The Heart

One effect of a treatment or mutation on female mice may be to alter the rate of the estrus cycle throughout the lifetime or experimental time course, as was the case for BRCA1−/− mice (Hong et al. 2010). It may be unfeasible, or experimentally undesirable, to measure the exact length of multiple, consecutive estrus cycles. Furthermore, investigators may be interested in indications that the treatment, condition, or mutation that they applied onto their mice altered circulating estrogen levels beyond the effect of normal estrus. This question is relevant when studying endocrine disrupting agents or steroid hormones, which often interact in physiological feedback loops. In the cases in which estrogen levels or endocrine disrupting agent levels were not directly measured in blood plasma, the weight and fibrotic state of the heart may give clues to altered lifelong levels of circulating estrogen. Abundant evidence supports the role of circulating estrogen in preventing cardiac hypertrophy, cardiac thinning, and cardiac fibrosis. In particular, the estrogen receptor-beta (ER-b) is the protein that mediates this protective effect of estrogen. In female mice, Angiotensin II (AngII) causes cardiac hypertrophy and collagen deposition, which is inhibited by activated ER-b (Pedram et al. 2008). AngII was also shown to stimulate cardiac fibroblasts to become cardiac myofibroblasts by inducing the expression of TGFB1, which in turn induced expression of vimentin, fibronectin, and collagens I and II; all of which contribute to fibrosis. Treatment with estrogen or an ER-b agonist (dipropylnitrile) blocked all of these events (Pedram et al. 2010). The ER-b agonist β-LGND2 is also effective against AngII-induced cardiac pathology (Pedram et al. 2013).

The prediction of whether a treatment, condition, or mutation—if it affects life-long levels of circulating estrogen—should increase or decrease hypertrophy and fibrosis depends on the specifics each experiment. Nonetheless, cardiac weight can be measured with a scale and cardiac fibrosis can be measured by immuno-staining for collagen I, collagen II, vimentin, or fibronectin. These pieces of information will give clues about abnormal lifelong estrogenic activity. It is worth noting that as rodents and humans age, the heart naturally undergoes hypertrophy and fibrosis (Anversa et al. 1990; Cornwell et al. 1991; Olivetti et al. 1991; Swynghedauw et al. 1995), so age effects should be considered and adjusted for via appropriate control specimens. Lastly, the weight of the heart can be normalized by the length of the tibia, to adjust for the influence of differential bodily growth rate between genders or treatment groups (Stauffer et al. 2006).

Mammary Glands of Male Mice

Some strains of male mice maintain a small ductal structure in their mammary fat pads throughout adulthood. These male ductal structures do not expand like the female glands during puberty. Vandenberg and colleagues demonstrated that male mammary ducts can be induced to grow via exposure to bisphenol A (BPA), an endo-crine disrupting chemical and environmental pollutant (Vandenberg et al. 2013). The expansion of the male mammary ductal system is an excellent surrogate measure to determine if male mice within a treatment group were inadvertently exposed to estrogenic compounds or if a treatment of interest had a feminizing effect on males.

Also see the topic entitled "Genetic Background" (Sect. 1) for why collecting certain endocrine organs will be helpful for reconciling data or phenotypes from different mouse strains that underwent the same experimental procedures.

The Bimodal Distribution

Investigators often compare the average measurement between two treatment groups and use the student's t-test (a parametric test assuming a normal distribution of the data) or the Mann–Whitney test (a non-parametric test assuming non-normal distri-bution of the data). For these two tests, the scatter of the data points above and below the mean or median matters, especially for the t-test. Plotting data as scatter plots will reveal if the data are segregated into two groups. Having two groups of data points clustered apart from each other, separated by a few data points in between, suggests a bimodal distribution of the data, as opposed to a unimodal distribution where all data points clump together in one mass. Bimodal distributions may be the result of two distinct biological situations occurring within a treatment group.

Bimodal distributions can seem like normal "noise" or scatter in the data. But with a sufficient sample size and similar results across multiple, independent research groups, the treatment being studied may actually be interacting with distinct biological conditions present within the group of mice. The bimodal distribution should be kept in mind when studying a treatment that is not expected to yield a dramatic phenotype (i.e. a low dose of some biologically active agent). Low-dose or small-effect treatments may make small perturbations that, in concert with normal variations in physiological status, lead to eventually distinct phenotypes months to years later.

Also see the topic entitled "Appreciate the Estrus Cycle to Reduce Variation in Data" in Sect. 1 of this chapter for a discussion of how estrus staging can reveal bimodal distribution in data from female mice.

Circulating Tumors Cells Don't Stay on One Side of the Mouse

Laboratories that study reagents and techniques that track a cell or a molecule within a tumor often do this in rodents. A relatively common experimental procedure is to label cancer cells by attaching a fluorescent marker, then inject the labeled cells into only one side of the mouse, and then track the location of the marked cells via non-invasive imaging techniques. A caveat about conclusions from such studies is that injected cancer cells or endogenous tumors often have circulating tumor cells. These cells leave the tumor and travel throughout the body through the blood stream or lymphatic system. Thus, tumor cells on the left side of the mouse can and do travel to the other side, meaning they can inhabit the other tumor. Thus, the other side of the mouse is not a good location to inject unlabeled cells of the control group. Data resulting from bilaterally injected mice may be confounded by the fact that tumor cells can migrate. Also, researchers may find that the bone marrow or organs other than the tumor exhibits the presence of the tracer. This may be due to the fact that the circulating tumor cells have seeded or metastasized to distant sites rather than that the fluorescent probe was metabolized and stored at the distant sites.

The Non-ubiquitous Activation of a Ubiquitous Artificial Promoter

An artificial gene promoter may only be activated in a subset of cells in a tissue or organ that ubiquitously expresses the transgene. An example is the mouse mammary tumor virus (MMTV) promoter, which is a common gene promoter that is used to drive the artificial expression of genes within the mouse mammary gland. The MMTV promoter is responsive to glucocorticoids and progesterone (Bruggemeier et al. 1991; Truss et al. 1995), both of which are a class of steroid hormones that bind to and activate nuclear hormone receptors that then act as transcription factors. Thus, cells that express the progesterone receptors and/or the glucocorticoid receptors will activate the MMTV promoter much more than cells that don't have these protein receptors. This becomes highly relevant for inducible knockout model systems in which the MMTV promoter controls the expression of Cre recombinase. Expression of the Cre gene by the MMTV promoter in a cell results in the excision of the target gene that is flanked by loxp sites in that same cell. The ductal epithelium of the mouse mammary gland contains two main layers of cells, the inner luminal epithelia and the outer myoepithelia. The transcriptional milieu between the two layers is distinct, reflecting their developmental origins from stem or progenitor cells: many of the luminal cells, as opposed to none of the myoepithelial cells, express the progesterone receptor, meaning the MMTV promoter is more often activated in the luminal compartment. Furthermore, lineage tracing and sorting studies have uncovered various types of progenitor cells that are luminal-like in nature (Shehata et al. 2012; Sleeman et al. 2007; van Amerongen et al. 2012; Van Keymeulen et al. 2011), meaning each has the potential to activate the MMTV promoter. No matter what tissue is being studied, it is good to know which cellular compartments are most likely to activate an artificial promoter. This information will undoubtedly shed light on mouse tumor data using artificial promoters.

Also see the topic entitled "Varied Mechanisms Can Yield the Same Phenotype: A Shrinking Tumor" in Sect. 3 for a discussion about variations in the spectrum of histological subtypes of tumors that yield the same phenotype of a shrinking or "cured" tumor. That type of variation can compound with the non-ubiquitous activation of artificial promoters in explaining incongruent mouse data that should be similar.

Varied Mechanisms Can Yield the Same Phenotype: A Shrinking Tumor

The golden read-out of studies on anti-cancer therapies is the tumor that shrinks due to treatment. Multiple biological mechanisms individually, but often in combination, can give rise to the phenotype of a shrinking tumor. The most common mechanisms are necrosis, apoptosis, senescence, dormancy, and immune clearance. The main mechanism by which a tumor shrinks may give clues to why two data sets derived from the same mouse model or xenografted mouse model yielded different results.

In certain cases when independent research groups have conflicting data about similar treatments or treatment modalities (i.e. dose, dosage patterns, priming dose followed by full dose), it is useful to examine the histological features of the tumors via hematoxylin & eosin (H&E) staining. For example, histology reveals not just the presence of necrosis, but the pattern of necrosis. Different histopathological subtypes (i.e. adenocarcinoma, squamous carcinoma, spindle cell carcinoma, etc.) can exhibit different patterns of the aforementioned mechanisms underlying tumor shrinkage. In the case of necrosis, the patterns can be a large, central necrotic core; a necrotic core along the tumor border; or many small necrotic cores throughout the tumor.

Aside from histology, simply keeping a record of the macroscopic features of tumors may yield useful information when for comparing incongruent data between experiments or research groups. Dissected tumors can range in color from an opaque whiteness to a dark reddish blackness. Their texture can range from solid and firm to soft and mushy. Tumors can spill out a viscous liquid from their middle region when dissected, be solid, be solid with dispersed pockets of viscous liquid, or bleed profusely. These macroscopic observations have different underlying biological mechanisms that give rise to them, which may provide crucial information for reconciling conflicting data sets that were expected to be similar.

The presence of different histologies and macroscopic characteristics between two data sets that employed the same xenografted human cell line can be very telling about why the data sets were incongruent. Excluding considerations of mislabeled cell lines, contaminated cell lines, and inadvertent "sub-cloning" of a cell line from the original cell line, incongruent macroscopic characteristics may indicate that the investigators are comparing apples to oranges.

Tumors resulting from a specific transgene that is regulated by an artificial promoter may exhibit different histologies. This may be because the transgene or transgene system (i.e. Cre recombinase systems) was randomly activated within a population of somatic cells. Activation of the transgenic system in a differentiated somatic cell may give rise to a different tumor histology than would a somatic early progenitor or late progenitor. See also the topic entitled "The Non-ubiquitous Activation of a Ubiquitous Artificial Promoter."

The Gut Microbiome Affects Cancer

The discovery that certain bacterial populations within the mammalian intestine can affect cancer has been very exciting, because it revealed yet another unrealized fundamental physiological mechanism of tumor development (reviewed in (Poutahidis et al. 2014; Walsh et al. 2014)). Furthermore, these discoveries suggest that there is future potential to develop non-invasive, dietary interventions that can inhibit or prevent cancer. In the context of data from mouse models of cancer, the gut microbiome immediately stands out as a confounding mechanism that has likely been the culprit of many irreproducible or incongruent mouse experiments. Our meager understanding of the gut microbiome's influence on cancer is both exciting and daunting. Like other fundamental mechanisms of physiology and cancer biology, it is a piece of the puzzle that is itself complex, yet is sure to interact with other known pieces of the puzzle in synergistically complicated ways.

See the topic entitled "Appreciate the Estrus Cycle to Reduce Variation in Data" in Sect. 1 of this chapter for a discussion about how hormones of the estrus cycle affect the immune system. The endocrinology of the estrus cycle will certainly be a factor that influences studies about interactions between the microbiome and cancer.

Due to space limitations, only three recent mouse studies about the intersection of the microbiome and cancer will be discussed here. They will suffice to show that someone needs to develop a simple, cost-effective method of preserving rodent fecal matter, or the microbes extracted from fecal matter, for future analyses. In this way, investigators can keep a record of viable gut microbiota from their mouse experiments as surrogate measures to reconcile conflicting or incongruent data sets. A consortium among research centers that agrees to process and store fecal samples from laboratories around the world may be an option.

The first study is entitled "The Gut Microbiome Modulates Colon Tumorigenesis" (Zackular et al. 2013). Zackular and colleagues studied the effect of altering the gut microbiome in a mouse model of inflammation-dependent colon cancer. They found that the proportions of different populations of bacteria changed as the mouse colon progressed from normal, to inflamed, to tumor-bearing. Mice that received an antibiotic cocktail to ablate their gut microbiome had much fewer and smaller colon tumors, showing a dramatic influence of the gut microbiome on cancer phenotype. Furthermore, the authors inoculated germ-free mice with the feces and bedding of tumor-bearing mice or healthy mice. The germ-free mice living in the infected bedding from tumor-bearing mice eventually harbored all phyla of bacteria and 90 % of genera of bacteria that were originally detected in the tumor-bearing mice. When colon tumors were induced via the same carcinogenic initiation & promotion chemicals as was used throughout the study, the germ-free mice living in infected bedding had twice as many tumors, and larger tumors, than the treated germ-free mice that were living in bedding from healthy mice.

The second study is entitled "The Intestinal Microbiota Modulates the Anticancer Immune Effects of Cyclophosphamide" (Viaud et al. 2013). Viaud and colleagues showed that the efficacy of cyclophosphamide (CTX), an anti-cancer agent, is dependent upon the presence of gram-positive bacteria within the small intestine of the mice (Viaud et al. 2013). The anti-cancer mechanism of CTX includes modulating the immune system into an anti-tumor phenotype. Viaud et al. showed that CTX caused damage to the lining of the small intestine, which allowed certain species of bacteria to translocate into the blood stream and mesenteric lymph nodes. Removal of bacteria via treatment with antibiotics ablated the anti-tumor efficacy of CTX. Furthermore, adoptive transfer of bacteria from CTX-treated mice to naïve mice activated T lymphocytes in the naïve mice into an anti-tumor phenotype.

The third study is entitled "Commensal Bacteria Control Cancer Response to Therapy by Modulating the Tumor Microenvironment" (Iida et al. 2013). Iida and colleagues studied the effectiveness of two classes of chemotherapy on tumors after ablating the gut microbiota via an antibiotic cocktail. One therapy was treatment with CpG-oligodeoxynucleotides along with an anti-IL-10 receptor antibody (anti-IL-10R/CpG-ODN). This treatment is known to induce necrosis in the MC38 tumor cells that were growing in the mouse. It does so by inducing cytokine production from myeloid cells of the innate immune system against the tumor, followed by activation of T lymphocytes against the tumor. However, in mice with ablated gut microbiota, the anti-IL-10R/CpG-ODN treatment induced less necrosis and less tumor shrinkage. This impaired reduction was also observed in *Rag1*−/− mice, which lack mature T and B lymphocytes. Together, these data suggest that the gut microbiota primarily interacts with the innate immune system as opposed to the adaptive immune system in mediating the anti-cancer effects of anti-IL-10R/CpG-ODN. The authors also showed that ablation of the gut microbiota also inhibited the effectiveness of two platinum-based chemotherapy compounds, one of which can kill tumor cells directly, without activating the immune system first.

In summary, the above three studies emphatically underscore the importance of the gut microbiome in modulating cancer development and chemotherapy response. Each study presents more questions than it provides answers, meaning much remains to be discovered.

References

Anversa, P., Palackal, T., Sonnenblick, E. H., Olivetti, G., Meggs, L. G., & Capasso, J. M. (1990). Myocyte cell loss and myocyte cellular hyperplasia in the hypertrophied aging rat heart. *Circulation Research, 67*, 871–885.

Aukema, H. M., Housini, I., & Rawling, J. M. (1999). Dietary soy protein effects on inherited polycystic kidney disease are influenced by gender and protein level. *Journal of the American Society of Nephrology, 10*, 300–308.

Ayakawa, S., Shibamoto, Y., Sugie, C., Ito, M., Ogino, H., Tomita, N., et al. (2009). Antitumor effects of a cyclooxygenase-2 inhibitor, meloxicam, alone and in combination with radiation and/or 5-fluorouracil in cultured tumor cells. *Molecular Medicine Reports, 2*, 621–625.

Beagley, K. W., & Gockel, C. M. (2003). Regulation of innate and adaptive immunity by the female sex hormones oestradiol and progesterone. *FEMS Immunology and Medical Microbiology, 38*, 13–22.

Benaroya-Milshtein, N., Apter, A., Yaniv, I., Kukulansky, T., Raz, N., Haberman, Y., et al. (2007). Environmental enrichment augments the efficacy of idiotype vaccination for B-cell lymphoma. *Journal of Immunotherapy, 30*, 517–522.

Brisken, C. (2013). Progesterone signalling in breast cancer: A neglected hormone coming into the limelight. *Nature Reviews Cancer, 13*, 385–396.

Bronte, V., & Pittet, M. J. (2013). The spleen in local and systemic regulation of immunity. *Immunity, 39*, 806–818.

Bruggemeier, U., Kalff, M., Franke, S., Scheidereit, C., & Beato, M. (1991). Ubiquitous transcription factor OTF-1 mediates induction of the MMTV promoter through synergistic interaction with hormone receptors. *Cell, 64*, 565–572.

Burkholder, T., Foltz, C., Karlsson, E., Linton, C. G., & Smith, J. M. (2012). Health evaluation of experimental laboratory mice. *Current Protocols in Mouse Biology, 2*, 145–165.

Byers, S. L., Wiles, M. V., Dunn, S. L., & Taft, R. A. (2012). Mouse estrous cycle identification tool and images. *PLoS One, 7*, e35538.

Cao, L., Liu, X., Lin, E. J., Wang, C., Choi, E. Y., Riban, V., et al. (2010). Environmental and genetic activation of a brain-adipocyte BDNF/leptin axis causes cancer remission and inhibition. *Cell, 142*, 52–64.

Chua, A. C., Hodson, L. J., Moldenhauer, L. M., Robertson, S. A., & Ingman, W. V. (2010). Dual roles for macrophages in ovarian cycle-associated development and remodelling of the mammary gland epithelium. *Development, 137*, 4229–4238.

Cornwell, G. G., 3rd, Thomas, B. P., & Snyder, D. L. (1991). Myocardial fibrosis in aging germ-free and conventional Lobund-Wistar rats: The protective effect of diet restriction. *Journal of Gerontology, 46*, B167–B170.

Davie, S. A., Maglione, J. E., Manner, C. K., Young, D., Cardiff, R. D., MacLeod, C. L., et al. (2007). Effects of FVB/NJ and C57Bl/6J strain backgrounds on mammary tumor phenotype in inducible nitric oxide synthase deficient mice. *Transgenic Research, 16*, 193–201.

Degen, G. H., Janning, P., Diel, P., & Bolt, H. M. (2002). Estrogenic isoflavones in rodent diets. *Toxicology Letters, 128*, 145–157.

Dominguez-Gerpe, L., & Rey-Mendez, M. (2003). Evolution of the thymus size in response to physiological and random events throughout life. *Microscopy Research and Technique, 62*, 464–476.

Donehower, L. A. (1996). The p53-deficient mouse: A model for basic and applied cancer studies. *Seminars in Cancer Biology, 7*, 269–278.

Donehower, L. A., Harvey, M., Slagle, B. L., McArthur, M. J., Montgomery, C. A., Jr., Butel, J. S., et al. (1992). Mice deficient for p53 are developmentally normal but susceptible to spontaneous tumours. *Nature, 356*, 215–221.

Ducottet, C., Aubert, A., & Belzung, C. (2004). Susceptibility to subchronic unpredictable stress is related to individual reactivity to threat stimuli in mice. *Behavioural Brain Research, 155*, 291–299.

Ducottet, C., & Belzung, C. (2004). Behaviour in the elevated plus-maze predicts coping after subchronic mild stress in mice. *Physiology & Behavior, 81*, 417–426.

Freund, R., Dubensky, T., Bronson, R., Sotnikov, A., Carroll, J., & Benjamin, T. (1992). Polyoma tumorigenesis in mice: Evidence for dominant resistance and dominant susceptibility genes of the host. *Virology, 191*, 724–731.

Goldman, A. P., Williams, C. S., Sheng, H., Lamps, L. W., Williams, V. P., Pairet, M., et al. (1998). Meloxicam inhibits the growth of colorectal cancer cells. *Carcinogenesis, 19*, 2195–2199.

Griffiths, J., Shanks, N., & Anisman, H. (1992). Strain-specific alterations in consumption of a palatable diet following repeated stressor exposure. *Pharmacology, Biochemistry, and Behavior, 42*, 219–227.

Hayden, E. C. (2014). Misleading mouse studies waste medical resources. *Nature News* (Nature Publishing Group).

Hong, H., Yen, H. Y., Brockmeyer, A., Liu, Y., Chodankar, R., Pike, M. C., et al. (2010). Changes in the mouse estrus cycle in response to BRCA1 inactivation suggest a potential link between risk factors for familial and sporadic ovarian cancer. *Cancer Research, 70*, 221–228.

Hooijmans, C. R., & Ritskes-Hoitinga, M. (2013). Progress in using systematic reviews of animal studies to improve translational research. *PLoS Medicine, 10*, e1001482.

Huitink, J. M., Heimerikxs, M., Nieuwland, M., Loer, S. A., Brugman, W., Velds, A., et al. (2010). Volatile anesthetics modulate gene expression in breast and brain tumor cells. *Anesthesia and Analgesia, 111*, 1411–1415.

IACUC. (2011). *Anesthesia and analgesia in laboratory animals at UCSF*. San Francisco: University of California.

Ibarguen-Vargas, Y., Surget, A., Touma, C., Palme, R., & Belzung, C. (2008). Multifaceted strain-specific effects in a mouse model of depression and of antidepressant reversal. *Psychoneuroendocrinology, 33*, 1357–1368.

Iida, N., Dzutsev, A., Stewart, C. A., Smith, L., Bouladoux, N., Weingarten, R. A., et al. (2013). Commensal bacteria control cancer response to therapy by modulating the tumor microenvironment. *Science, 342*, 967–970.

Jerry, D. J., Kittrell, F. S., Kuperwasser, C., Laucirica, R., Dickinson, E. S., Bonilla, P. J., et al. (2000). A mammary-specific model demonstrates the role of the p53 tumor suppressor gene in tumor development. *Oncogene, 19*, 1052–1058.

Kilkenny, C., Browne, W. J., Cuthill, I. C., Emerson, M., & Altman, D. G. (2010). Improving bioscience research reporting: The ARRIVE guidelines for reporting animal research. *PLoS Biology, 8*, e1000412.

Kuperwasser, C., Hurlbut, G. D., Kittrell, F. S., Dickinson, E. S., Laucirica, R., Medina, D., et al. (2000). Development of spontaneous mammary tumors in BALB/c p53 heterozygous mice. A model for Li-Fraumeni syndrome. *The American Journal of Pathology, 157*, 2151–2159.

Lewis, L. A., Huskey, P. S., & Kusewitt, D. F. (2012). High incidence of scrotal hernias in a closed colony of FVB mice. *Comparative Medicine, 62*, 391–394.

Lutz, D. (2011). *New study calls into question reliance on animal models in cardiovascular research*. St. Louis: Washington Univesity in St Louis Newsroom.

MacCord, K. (2013). *Germ layers*. Embryo Project Encyclopedia.

Maronpot, R. R., Yoshizawa, K., Nyska, A., Harada, T., Flake, G., Mueller, G., et al. (2010). Hepatic enzyme induction: Histopathology. *Toxicologic Pathology, 38*, 776–795.

Meijer, M. K., Sommer, R., Spruijt, B. M., van Zutphen, L. F., & Baumans, V. (2007). Influence of environmental enrichment and handling on the acute stress response in individually housed mice. *Laboratory Animals, 41*, 161–173.

Mineur, Y. S., Belzung, C., & Crusio, W. E. (2006). Effects of unpredictable chronic mild stress on anxiety and depression-like behavior in mice. *Behavioural Brain Research, 175*, 43–50.

Muhlhausler, B. S., Bloomfield, F. H., & Gillman, M. W. (2013). Whole animal experiments should be more like human randomized controlled trials. *PLoS Biology, 11*, e1001481.

Nachat-Kappes, R., Pinel, A., Combe, K., Lamas, B., Farges, M. C., Rossary, A., et al. (2012). Effects of enriched environment on COX-2, leptin and eicosanoids in a mouse model of breast cancer. *PLoS One, 7*, e51525.

Oakes, S. R., Hilton, H. N., & Ormandy, C. J. (2006). The alveolar switch: Coordinating the proliferative cues and cell fate decisions that drive the formation of lobuloalveoli from ductal epithelium. *Breast Cancer Research, 8*, 207.

Olivetti, G., Melissari, M., Capasso, J. M., & Anversa, P. (1991). Cardiomyopathy of the aging human heart. Myocyte loss and reactive cellular hypertrophy. *Circulation Research, 68*, 1560–1568.

Pearse, G. (2006). Normal structure, function and histology of the thymus. *Toxicologic Pathology, 34*, 504–514.

Pedram, A., Razandi, M., Korach, K. S., Narayanan, R., Dalton, J. T., & Levin, E. R. (2013). ERbeta selective agonist inhibits angiotensin-induced cardiovascular pathology in female mice. *Endocrinology, 154*, 4352–4364.

Pedram, A., Razandi, M., Lubahn, D., Liu, J., Vannan, M., & Levin, E. R. (2008). Estrogen inhibits cardiac hypertrophy: Role of estrogen receptor-beta to inhibit calcineurin. *Endocrinology, 149*, 3361–3369.

Pedram, A., Razandi, M., O'Mahony, F., Lubahn, D., & Levin, E. R. (2010). Estrogen receptor-beta prevents cardiac fibrosis. *Molecular Endocrinology, 24*, 2152–2165.

Petrovska, M., Dimitrov, D. G., & Michael, S. D. (1996). Quantitative changes in macrophage distribution in normal mouse ovary over the course of the estrous cycle examined with an image analysis system. *American Journal of Reproductive Immunology, 36*, 175–183.

Plante, I., Stewart, M. K., & Laird, D. W. (2011). Evaluation of mammary gland development and function in mouse models. *Journal of Visualized Experiments, 53*, e2828.

Pothion, S., Bizot, J. C., Trovero, F., & Belzung, C. (2004). Strain differences in sucrose preference and in the consequences of unpredictable chronic mild stress. *Behavioural Brain Research, 155*, 135–146.

Pound, P., Ebrahim, S., Sandercock, P., Bracken, M. B., Roberts, I., & Reviewing Animal Trials Systematically Group. (2004). Where is the evidence that animal research benefits humans? *BMJ, 328*, 514–517.

Poutahidis, T., Kleinewietfeld, M., & Erdman, S. E. (2014). Gut microbiota and the paradox of cancer immunotherapy. *Frontiers in Immunology, 5*, 157.

Quiner, T. E., Nakken, H. L., Mason, B. A., Lephart, E. D., Hancock, C. R., & Christensen, M. J. (2011). Soy content of basal diets determines the effects of supplemental selenium in male mice. *The Journal of Nutrition, 141*, 2159–2165.

Reinhardt, H. C., & Schumacher, B. (2012). The p53 network: Cellular and systemic DNA damage responses in aging and cancer. *Trends in Genetics, 28*, 128–136.

SABRE-Research-UK (2014). Systematic reviews: Quality of evidence pyramid. http://www.sabre.org.uk/systematic-reviews/4578536825.

Sahar, S., & Sassone-Corsi, P. (2009). Metabolism and cancer: The circadian clock connection. *Nature Reviews Cancer, 9*, 886–896.

Savvidis, C., & Koutsilieris, M. (2012). Circadian rhythm disruption in cancer biology. *Molecular Medicine, 18*, 1249–1260.

Sena, E. S., Currie, G. L., McCann, S. K., Macleod, M. R., & Howells, D. W. (2014). Systematic reviews and meta-analysis of preclinical studies: Why perform them and how to appraise them critically. *Journal of Cerebral Blood Flow and Metabolism, 34*, 737–742.

Shehata, M., Teschendorff, A., Sharp, G., Novcic, N., Russell, I. A., Avril, S., et al. (2012). Phenotypic and functional characterisation of the luminal cell hierarchy of the mammary gland. *Breast Cancer Research, 14*, R134.

Sleeman, K. E., Kendrick, H., Robertson, D., Isacke, C. M., Ashworth, A., & Smalley, M. J. (2007). Dissociation of estrogen receptor expression and in vivo stem cell activity in the mammary gland. *The Journal of Cell Biology, 176*, 19–26. doi:10.1083/jcb.200604065.

Sorge, R. E., Martin, L. J., Isbester, K. A., Sotocinal, S. G., Rosen, S., Tuttle, A. H., et al. (2014). Olfactory exposure to males, including men, causes stress and related analgesia in rodents. *Nature Methods, 11*, 629–632.

Spike, B. T., & Wahl, G. M. (2011). p53, stem cells, and reprogramming: Tumor suppression beyond guarding the genome. *Genes & Cancer, 2*, 404–419.

Stauffer, B. L., Konhilas, J. P., Luczak, E. D., & Leinwand, L. A. (2006). Soy diet worsens heart disease in mice. *The Journal of Clinical Investigation, 116*, 209–216.

Sternlicht, M. D. (2006). Key stages in mammary gland development: The cues that regulate ductal branching morphogenesis. *Breast Cancer Research, 8*, 201.

Stutman, O. (1974). Cell-mediated immunity and aging. *Federation Proceedings, 33*, 2028–2032.

Swynghedauw, B., Besse, S., Assayag, P., Carre, F., Chevalier, B., Charlemagne, D., et al. (1995). Molecular and cellular biology of the senescent hypertrophied and failing heart. *The American Journal of Cardiology, 76*, 2D–7D.

Takahashi, J. S., Hong, H. K., Ko, C. H., & McDearmon, E. L. (2008). The genetics of mammalian circadian order and disorder: Implications for physiology and disease. *Nature Reviews Genetics, 9*, 764–775.

Tam, P. P., & Beddington, R. S. (1992). Establishment and organization of germ layers in the gastrulating mouse embryo. *Ciba Foundation Symposium, 165*, 27–41. discussion 42–49.

Thigpen, J. E., Setchell, K. D., Saunders, H. E., Haseman, J. K., Grant, M. G., & Forsythe, D. B. (2004). Selecting the appropriate rodent diet for endocrine disruptor research and testing studies. *ILAR Journal, 45*, 401–416.

Truss, M., Bartsch, J., Schelbert, A., Hache, R. J., & Beato, M. (1995). Hormone induces binding of receptors and transcription factors to a rearranged nucleosome on the MMTV promoter in vivo. *The EMBO Journal, 14*, 1737–1751.

van Amerongen, R., Bowman, A. N., & Nusse, R. (2012). Developmental stage and time dictate the fate of Wnt/beta-catenin-responsive stem cells in the mammary gland. *Cell Stem Cell, 11*, 387–400.

Van Keymeulen, A., Rocha, A. S., Ousset, M., Beck, B., Bouvencourt, G., Rock, J., et al. (2011). Distinct stem cells contribute to mammary gland development and maintenance. *Nature, 479*, 189–193.

Vandenberg, L. N., Schaeberle, C. M., Rubin, B. S., Sonnenschein, C., & Soto, A. M. (2013). The male mammary gland: A target for the xenoestrogen bisphenol A. *Reproductive Toxicology, 37*, 15–23.

Viaud, S., Saccheri, F., Mignot, G., Yamazaki, T., Daillere, R., Hannani, D., et al. (2013). The intestinal microbiota modulates the anticancer immune effects of cyclophosphamide. *Science, 342*, 971–976.

Vousden, K. H., & Ryan, K. M. (2009). p53 and metabolism. *Nature Reviews Cancer, 9*, 691–700.

Walsh, C. J., Guinane, C. M., O'Toole, P. W., & Cotter, P. D. (2014). Beneficial modulation of the gut microbiota. *FEBS Letters, 588*(22), 4120–4130.

Westwood, J. A., Darcy, P. K., & Kershaw, M. H. (2013). Environmental enrichment does not impact on tumor growth in mice. *F1000Research, 2*, 140.

Xin, B., Yokoyama, Y., Shigeto, T., Futagami, M., & Mizunuma, H. (2007). Inhibitory effect of meloxicam, a selective cyclooxygenase-2 inhibitor, and ciglitazone, a peroxisome proliferator-activated receptor gamma ligand, on the growth of human ovarian cancers. *Cancer, 110*, 791–800.

Zackular, J. P., Baxter, N. T., Iverson, K. D., Sadler, W. D., Petrosino, J. F., Chen, G. Y., et al. (2013). The gut microbiome modulates colon tumorigenesis. *mBio, 4*, e00692–00613.

Index

A
Anti-IL-10 receptor antibody (anti-IL-10R/
  CpG-ODN), 50

B
Bisphenol A (BPA), 5
Breast cancer, 14
  ECM (*see* Extracellular matrix (ECM))
  epidemiology, 2
  estrogens
    EDCs, 4–5
    ER-α and ER-β, 4
    ROS production, 6–7
  estrus cycle
    mammary gland, 2–3
    ovarian function, 3–4
  systems biology perspective, 1

C
Cancer stem cell (CSC)
  cell fusion, 29
  clonality, 26
  dedifferentiation, 24
  differentiation, 23
  DNA and protein damage, 26
  EMT, 24, 25, 29–30
  function and identity, 27
  lineage commitment, 23
  microenvironment, 25, 27
  niche, 25
  plasticity, 24, 28
  potency, 23
  progenitor cell, 22

properties, 22
transdifferentiation, 24
Cells-of-origin. *See also* Cancer stem cell
  (CSC)
  function and identity, 27
  hematopoietic cancers, 28
  microenvironment, 27
  vs. plasticity, 29–30
  progenitor cells, 27
Cellular plasticity, 28
Cyclophosphamide (CTX), 50

E
Endocrine disrupting chemicals (EDCs), 4–5
Endocrine systems
  heart, 45–46
  liver, 45
  mammary glands, 46
  spleen, 44
  surrogate metrics, 44
  thymus, 44
Epithelial-to-mesenchymal transition (EMT),
  24, 25, 29–30
Estrus cycle
  mammary gland, 2–3
  mouse models
    daily staging, 36–37
    immune systems, 36
    proestrus, metestrus, and diestrus, 35
    stage identification, 36
  ovaries, 3–4
Extracellular matrix (ECM)
  composition, 7
  directly testing, 11–12

© Springer International Publishing Switzerland 2016
D.H. Nguyen, *Systems Biology of Tumor Physiology*, SpringerBriefs
in Cancer Research, DOI 10.1007/978-3-319-25601-6

Extracellular matrix (ECM) (*cont.*)
  indirectly testing
    aging, 12–13
    POF, 13–14
    post-menopausal women, 14
    WAS, 14–15
  lysyl oxidase, 9
  macrophage phenotypes, 9–10
  mammary gland, 7–8
  PTEN, 9

H
Hormone replacement therapy, 14

M
Macrophages. *See* Breast cancer
Mouse mammary tumor virus (MMTV)
    promoter, 47–48
Mouse models, 35–37, 44–46
  age of, 35
  anesthesia, 42–43
  bimodal distributions, 46–47
  composition, 39–40
  deformity, 39
  endocrine systems
    heart, 45–46
    liver, 45
    mammary glands, 46
    spleen, 44
    surrogate metrics, 44
    thymus, 44
  environmental enrichment, 41

estrus cycle
  daily staging, 36–37
  immune systems, 36
  proestrus, metestrus, and diestrus, 35
  stage identification, 36
fluorescent marker, 47
frustration, 34
genetics, 37–38
gut microbiome, 49–50
histological features, 48
macroscopic features, 48
mechanisms, 48
non-ubiquitous activation, 47–48
rodent models, 33
stressors, 40
transgene system, 49
treatments and exposures, 42

P
Phosphatase and tensin homolog (PTEN), 9
Premature ovarian failure (POF), 13–14

R
Reactive oxygen species (ROS), 6–7

S
Single nucleotide polymorphism (SNP), 38

W
Wiskott–Aldrich syndrome (WAS), 14–15

Printed in the United States
By Bookmasters